£21.99

BTEC National
FURTHER
MATHEMATICS
for Technicians

3rd edition

3rd edition

BTEC National FURTHER MATHEMATICS for Technicians

Graham Taylor
Alex Greer

First published in 1982 by:
Stanley Thornes (Publishers) Ltd
Second edition 1989

Third edition published in 2005 by:
Nelson Thornes Ltd
Delta Place
27 Bath Road
CHELTENHAM
GL53 7TH
United Kingdom

08 09 / 10 9 8 7 6 5 4 3 2

A catalogue record for this book is available from the British Library

ISBN 978 0 7487 9410 2

Page make-up by Tech-Set Ltd, Gateshead, Tyne and Wear

Printed and bound in Spain by GraphyCems

Contents

About this Book

The Specifications

This book covers Further Mathematics for Technicians, unit 32, at BTEC National NVQ level 3. This unit is used for students following either a mechanical or electrical path in engineering.

The book broadly follows unit structure learning outcomes. These are all application techniques to the solution of engineering problems: Algebraic, trigonometry, calculus, probability and statistics.

Readers will welcome extensive revision, and in many topics no previous knowledge is assumed.

Assessment

BTEC qualifications are awarded in three grades:

Pass　　　 This shows that you have a basic knowledge and understanding,

Merit　　　 This shows that you have a sound knowledge and understanding,

Distinction　 This requires that you have an in-depth knowledge and understanding.

Authors' note

We have tried to follow a sympathetic approach to basic maths as required by Engineers. It is a book written for Engineers by Engineers, so we appreciate the approach needed to make the work practical and interesting.

We hope you will enjoy using this book, and find it a useful reference after you have finished your studies.

Non-linear Laws Reducible to Linear Form

1

Revision of the straight line – graphs directly reducible e.g. $y = ax^2 + b$, $y = a/x + b$ – logarithmic scales and graph paper – graphs reducible using log form e.g. $y = at^n$.

INTRODUCTION

Technologists are often faced with analysing a set of readings more often than not obtained from experimental work. We need to know whether there is any mathematical relationship between the pairs of values. One way is to plot them as a graph and/or analyse the results on a computer. However, before we resort to this we can try to arrange the values to give a straight line – this would be our dream as it enables easy analysis.

In this chapter you will see how we can try reducing relationships to straight lines – even changing the ball game by using logarithmic scales. It is unusual that we are completely in the dark since previous work history may well guide us swiftly to a result. We can then guess a suitable relationship and test if it is satisfactory.

We shall start by remembering how to plot a good graph by careful choice of scales and positioning of graph on the paper – also the standard equation of a straight line.

Axes and Scales

By good positioning of the axes on the graph paper and careful choice of scales (which need not be the same on each axis) the easier it will be to plot points and draw an accurate graph. To start with you may find this rather tedious, but with practice everything falls into place.

You may appreciate some advice on the choice of a scale on an axis. The scale should be as large as possible so that the points may be plotted with the greatest possible accuracy – this means that you should make the plot cover the whole sheet of graph paper, not just a small area in the bottom left-hand corner.

The difference between a good scale and a bad scale is how easy it is to read intermediate decimal values: try 1.7 on the 'bad' scales shown below and you will see why they why they are so named.

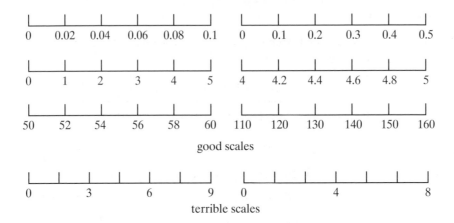

good scales

terrible scales

THE STRAIGHT LINE

The standard equation of the straight line: $\boxed{y = mx + c}$

If the origin O (the point 0, 0) is present at the intersection of the axes the values of gradient m and intercept c may be found directly from the graph.

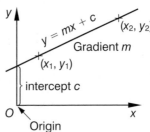

For graphs where the origin is not shown we choose two points (x_1, y_1) and (x_2, y_2) which both lie on the line, and substitute their values into the standard linear equation giving the simultaneous equations:

$$\left. \begin{array}{l} y_1 = mx_1 + c \\ y_2 = mx_2 + c \end{array} \right\} \begin{array}{l} \text{from which } m \text{ and } c \\ \text{may be found} \end{array}$$

Incidentally this also works if the origin **is** present.

The form $\quad y = \dfrac{a}{x} + b$

If we have pairs of (x, y) values which satisfy the given equation $y = a\left(\dfrac{1}{x}\right) + b$ then a plot of y vertically and $\dfrac{1}{x}$ horizontally will give a straight line.

EXAMPLE 1.1

An experiment connected with the flow of water over a rectangular weir gave the following results:

C	0.503	0.454	0.438	0.430	0.425	0.421
H	0.1	0.2	0.3	0.4	0.5	0.6

The relation between C and H is thought to be of the form $C = \dfrac{a}{H} + b$.

Test if this is so and find the values of the constants a and b.

In the suggested equation C is the sum of two terms, the first of which varies as $\dfrac{1}{H}$. If the equation $C = \dfrac{a}{H} + b$ is correct then when we plot

C against $\dfrac{1}{H}$ we should obtain a straight line. To do this we draw up the following table:

C	0.503	0.454	0.438	0.430	0.425	0.421
$\dfrac{1}{H}$	10.00	5.00	3.33	2.50	2.00	1.67

The graph obtained is shown in Fig. 1.1. It is a straight line and hence the given values follow a law of the form $C = \dfrac{a}{H} + b$.

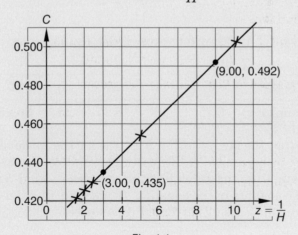

Fig. 1.1

To find the values of a and b we choose two points which *lie on the straight line*.

The point $(3.00, 0.435)$ lies on the line.

$$\therefore \qquad\qquad\qquad 0.435 = 3.00a + b \qquad\qquad\qquad [1]$$

The point $(9.00, 0.492)$ also lies on the line.

$$\therefore \qquad\qquad\qquad 0.492 = 9.00a + b \qquad\qquad\qquad [2]$$

Subtracting equation [1] from equation [2] gives

$$0.492 - 0.435 = a(9.00 - 3.00)$$
$$\therefore \qquad\qquad a = 0.0095$$

Substituting this value for a in equation [1] gives

$$0.435 = 3.00 \times 0.0095 + b$$
$$\therefore \qquad b = 0.435 - 0.0285 = 0.407$$

Hence the values of a and b are 0.0095 and 0.407, respectively.

The form $y = a(x^2) + b$

If we have pairs of (x, y) values which satisfy the given equation $y = a(x^2) + b$ then a plot of y vertically and x^2 horizontally will give a straight line.

EXAMPLE 1.2

The fusing current I amperes for wires of various diameters d mm is as shown below:

d (mm)	5	10	15	20	25
I (amperes)	6.25	10	16.25	25	36.25

It is suggested that the law $I = ad^2 + b$ is true for the range of values given, a and b being constants. By plotting a suitable graph show that this law holds and from the graph find the constants a and b. Using the values of these constants in the equation $I = ad^2 + b$ find the diameter of the wire required for a fusing current of 12 amperes.

By putting $z = d^2$ the equation $I = ad^2 + b$ becomes $I = az + b$ which is the standard form of a straight line. Hence by plotting I against d^2 we should get a straight line if the law is true. To try this we draw up a table showing corresponding values of I and d^2.

d	5	10	15	20	25
$z = d^2$	25	100	225	400	625
I	6.25	10	16.25	25	36.25

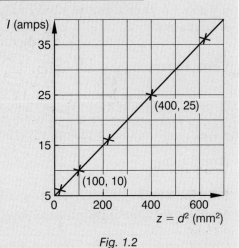

Fig. 1.2

From the graph (Fig. 1.2) we see that the points do lie on a straight line, and hence the values obey a law of the form $I = ad^2 + b$.

To find the values of constants a and b we choose two points which lie on the line and substitute their co-ordinates into the equation.

The point $(400, 25)$ lies on the line.

$\therefore$ $$25 = 400a + b \tag{1}$$

The point $(100, 10)$ lies on the line.

$\therefore$ $$10 = 100a + b \tag{2}$$

Subtracting equation [2] from equation [1] gives

$$15 = 300a$$
$\therefore$ $$a = 0.05$$

Substituting $a = 0.05$ in equation [2] gives

$$10 = 100 \times 0.05 + b$$
$\therefore$ $$b = 5$$

Therefore the law is $I = 0.05d^2 + 5$

When $I = 12$ $12 = 0.05d^2 + 5$

$\therefore$ $d = \sqrt{140} = 11.8\,\text{mm}$

The form $y = \dfrac{a}{x^2} + b$

If we have pairs of (x, y) values which satisfy the given equation $y = a\left(\dfrac{1}{x^2}\right) + b$ then a plot of y vertically and $\dfrac{1}{x^2}$ horizontally will give a straight line.

The form $y = a\sqrt{x} + b$

If we have pairs of (x, y) values which satisfy the given equation $y = a\sqrt{x} + b$ then a plot of y vertically and $\sqrt{x}$ horizontally will give a straight line.

The form $y = \dfrac{a}{\sqrt{x}} + b$

If we have pairs of (x, y) values which satisfy the given equation $y = a\left(\dfrac{1}{\sqrt{x}}\right) + b$ then a plot of y vertically and $\dfrac{1}{\sqrt{x}}$ horizontally will give a straight line.

Exercise 1.1

1) The following readings were taken during a test:

R (ohms)	85	73.3	64	58.8	55.8
I (amperes)	2	3	5	8	12

R and I are thought to be connected by an equation of the form $R = \dfrac{a}{I} + b$. Verify that this is so by plotting R (y-axis) against $\dfrac{1}{I}$ (x-axis) and hence find values for a and b.

2) In the theory of the moisture content of thermal insulation efficiency of porous materials the following table gives values of μ the diffusion constant of the material and k_m the thermal conductivity of damp insulation material:

μ	1.3	2.7	3.8	5.4	7.2	10.0
k_m	0.0336	0.0245	0.0221	0.0203	0.0192	0.0183

Find the equation connecting μ and k_m if it is of the form $k_m = a + \dfrac{b}{\mu}$ where a and b are constants.

3) The accompanying table gives the corresponding values of the pressure, p, of mercury and the volume, v, of a given mass of gas at constant temperature.

p	90	100	130	150	170	190
v	16.66	13.64	11.54	9.95	8.82	7.89

By plotting p against the reciprocal of v obtain some relation between p and v.

4) In an experiment, the resistance R of copper wire of various diameters d mm was measured and the following readings were obtained.

d (mm)	0.1	0.2	0.3	0.4	0.5
R (ohms)	20	5	2.2	1.3	0.8

Show that $R = \dfrac{k}{d^2}$ and find a suitable value for k.

5) The following table gives the thickness T mm of a brass flange brazed to a copper pipe of internal diameter D mm:

T (mm)	15.5	17.8	19.5	20.9	22.2	23.3
D (mm)	50	100	150	200	250	300

Show that T and D are connected by an equation of the form $T = a\sqrt{D} + b$, find the values of constants a and b, and find the thickness of the flange for a 70 mm diameter pipe.

6) The table shows how the coefficient of friction, μ, between a belt and a pulley varies with the speed, v m/s, of the belt. By plotting a graph show that $\mu = m\sqrt{v} + c$ and find the values of constants m and c.

μ	0.26	0.29	0.32	0.35	0.38
v	2.22	5.00	8.89	13.89	20.00

7) Using the table below show that the values are in agreement with the law $y = \dfrac{m}{\sqrt{x}} + c$. Hence evaluate the constants m and c.

x	0.2	0.8	1.2	1.8	2.5	4.4
y	1.62	1.51	1.49	1.47	1.46	1.44

REDUCTION TO LOGARITHMIC FORM

Not all equations may be reduced to linear form by making a relatively straightforward substitution. Some types need special treatment and we approach these by using our knowledge of logarithms and then plotting the values, given in the particular problem, on special graph paper having a logarithmic scale along one, or both, of the axes.

Equations of the type $z = at^n$, $z = ab^t$ and $z = ae^{bt}$

In all the work which follows in this chapter the logarithms used will be to the base 10 and are denoted by 'lg'.

Consider the following relationships in which z and t are the variables, whilst a, b and n are constants.

$z = at^n$

Fig. 1.3

Now
$$z = at^n$$

and taking logs
$$\lg z = \lg(at^n)$$
$$= \lg t^n + \lg a$$
$$\lg z = n \lg t + \lg a$$

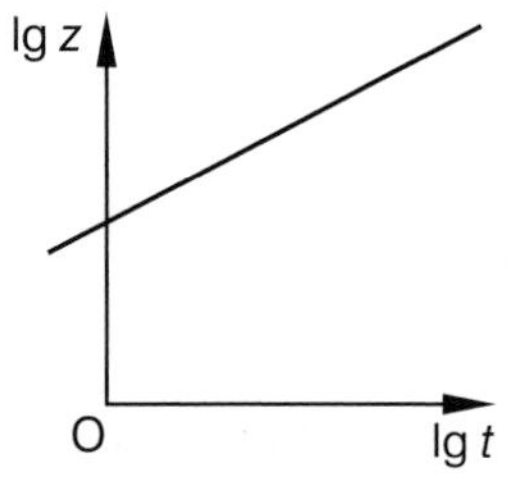

The given values of the variables will satisfy this equation if they satisfy the original equation. Comparing this equation with $y = mx + c$, which is the standard equation of a straight line, we see that if we plot $\lg z$ on the y-axis and $\lg t$ on the x-axis the result will be a straight line (Fig. 1.3) and the values of the constants n and a may be found using the two point method described earlier.

$z = ab^t$

Fig. 1.4

Now
$$z = ab^t$$

and taking logs
$$\lg z = \lg(ab^t)$$
$$= \lg b^t + \lg a$$
$$\lg z = (\lg b)t + \lg a$$

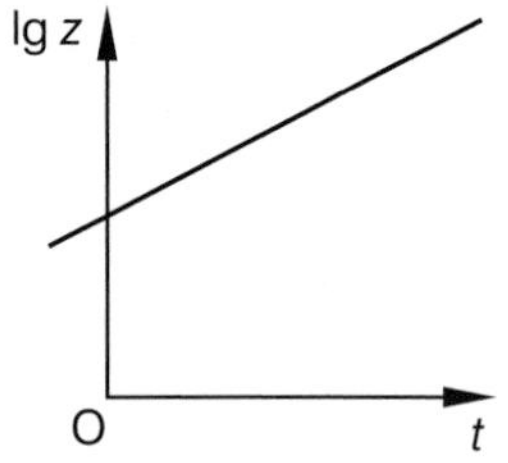

We now proceed in a manner similar to that used for the previous equation by plotting $\lg z$ on the y-axis and t on the x-axis, and again obtain a straight line (Fig. 1.4).

$z = ae^{bt}$

Now
$$z = ae^{bt}$$

and taking logs
$$\lg z = \lg(ae^{bt})$$
$$= \lg e^{bt} + \lg a$$
$$= (b \lg e)t + \lg a$$

but $\lg e = 0.4343$

$\therefore$
$$\lg z = (0.4343b)t + \lg a$$

Again proceeding in a manner similar to that used for the previous equation, we plot $\lg z$ on the y-axis and t on the x-axis and again obtain a straight line (Fig. 1.4).

Logarithmic Scales

In your previous work you may have used ordinary graph paper for problems involving logarithmic and exponent laws. This entailed finding the logs of each individual given number on at least one of the axes. However if logarithmic scales are used it is no longer necessary to find individual logs.

We use logs to the base 10 since, as you will see, natural logs to the base e would result in inconvenient numbers of the scales.

It has been shown earlier that: Number $=$ Base^{Logarithm}

and if we use a base of 10 then: Number $= 10^{\text{Logarithm}}$

Since	$1000 = 10^3$	then we may write	$\log_{10} 1000 = 3$
and since	$100 = 10^2$	then we may write	$\log_{10} 100 = 2$
and since	$100 = 10^1$	then we may write	$\log_{10} 10 = 1$
and since	$1 = 10^0$	then we may write	$\log_{10} 1 = 0$
and since	$0.1 = 10^{-1}$	then we may write	$\log_{10} 0.1 = -1$
and since	$0.01 = 10^{-2}$	then we may write	$\log_{10} 0.01 = -2$
and since	$0.001 = 10^{-3}$	then we may write	$\log_{10} 0.001 = -3$

and so on.

These logarithms may be shown on a scale as shown in Fig. 1.5.

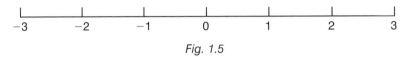

Fig. 1.5

However, since we wish to plot numbers directly on to the scale (without any reference to their logarithms) the scale is labelled as shown in Fig. 1.6.

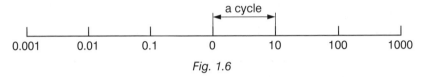

Fig. 1.6

Each division is called a cycle and is sub-divided using a logarithmic scale. Two such cycles are shown in Fig. 1.7.

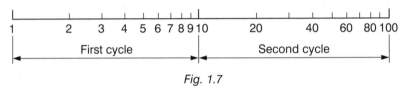

Fig. 1.7

The choice of numbers on the scale depends on the numbers allocated to the variables in the problem to be solved. Thus in Fig. 1.7 the numbers run from 1 to 100.

Logarithmic Graph Paper

Logarithmic scales may be used on graph paper in place of the more usual linear scales. By using graph paper ruled in this way log plots may be made without the necessity of looking up the logs of each given value. Semi-logarithmic graph paper is also available and has one way ruled with log scales whilst the other way has the usual linear scale. Examples of each are shown in Figs. 1.8 and 1.9.

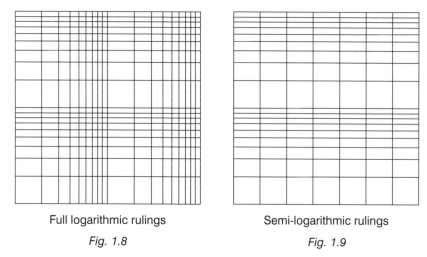

Full logarithmic rulings   Semi-logarithmic rulings

Fig. 1.8   *Fig. 1.9*

The use of these scales and the special graph paper is shown by the examples which follow.

EXAMPLE 1.3

The law connecting two quantities z and t is of the form $z = at^n$. Find the law given the following pairs of values:

z	3.170	4.603	7.499	10.50	15.17
t	7.980	9.863	13.03	15.81	19.50

The relationship $\quad\quad\quad\quad z = at^n$

gives (see text page 8) $\quad\quad \lg z = n \lg t + \lg a$ $\quad\quad\quad\quad$ [1]

Hence we plot the given values of z and t on log scales as shown in Fig. 1.10.

On both the vertical and horizontal axes we require 2 cycles, the first for values from 1 to 10 and the second for values from 10 to 100.

The constants are found by taking two pairs of cordinates:

Point $(25, 23)$ lies on the line, and putting these values in equation [1],

$$\lg 23 = n \lg 25 + \lg a \quad\quad\quad\quad [2]$$

Point $(4.1, 1)$ lies on the line and putting these values in equation [1],

$$\lg 1 = n \lg 4.1 + \lg a \quad\quad\quad\quad [3]$$

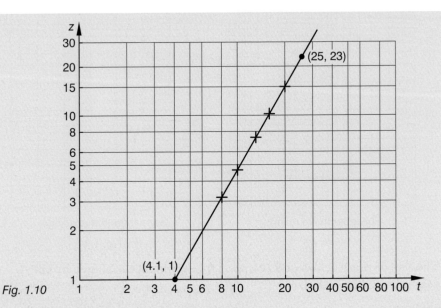

Fig. 1.10

Subtracting equation [3] from equation [2],

$$\lg 23 - \lg 1 = n(\lg 25 - \lg 4.1)$$
$$\lg(23/1) = n[\lg(25/4.1)]$$
$$n = \frac{\lg(23/1)}{\lg(25/4.1)}$$
$$n = 1.73$$

Substituting this value of n in equation [3],

$$\lg 1 = 1.73 \lg 4.1 + \lg a$$
$$\therefore \quad \lg a = \lg 1 - 1.73 \lg 4.1$$
$$\therefore \quad a = 0.087$$

Hence the law is $z = 0.087 t^{1.73}$.

EXAMPLE 1.4

The table gives values obtained in an experiment. It is thought that the law may be of the form $z = ab^t$, where a and b are constants. Verify this and find the law.

t	0.190	0.250	0.300	0.400
z	11 220	18 620	26 920	61 660

We think that the relationship is of the form:

$$z = ab^t$$

which gives (see text page 8),

$$\lg z = (\lg b)t + \lg a \tag{1}$$

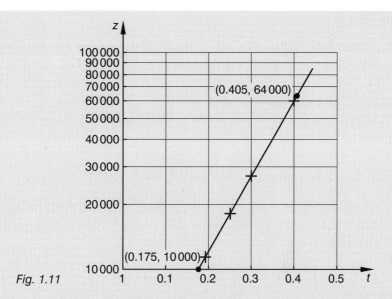

Fig. 1.11

Hence we plot the given values of z on a vertical log scale – the t values, however, will be on the horizontal axis on an ordinary linear scale (Fig. 1.11).

The points lie on a straight line, and hence the given values of z and t obey the law. We now have to find the cordinates of two points lying on the line.

Point $(0.405, 64\,000)$ lies on the line, and substituting in equation [1],

$$\lg 64\,000 = (\lg b)0.405 + \lg a \qquad [2]$$

Point $(0.175, 10\,000)$ lies on the line, and substituting in equation [1],

$$\lg 10\,000 = (\lg b)0.175 + \lg a \qquad [3]$$

Subtracting equation [3] from equation [2],

$$\lg 64\,000 - \lg 10\,000 = (\lg b)(0.405 - 0.175)$$
$$\therefore \qquad 4.8062 - 4.0000 = 0.230(\lg b)$$
$$\therefore \qquad \lg b = \frac{0.8062}{0.230} = 3.5052$$
$$b = 3200$$

Substituting in equation [3],

$$\lg 10\,000 = (3.5052)0.175 + \lg a$$
$$\therefore \qquad \lg a = \lg 10\,000 - 3.5052(0.175)$$
$$= 4 - 0.6134 = 3.3866$$
$$\therefore \qquad a = 2436$$

Hence the law is:

$$z = 2436(3200)^t$$

EXAMPLE 1.5

V and t are connected by the law $V = ae^{bt}$. If the values given in the table satisfy the law, find the constants a and b.

t	0.05	0.95	2.05	2.95
V	20.70	24.49	30.27	36.06

The law is: $$V = ae^{bt}$$

which gives (see text page 9), $\lg V = (0.4343b)t + \lg a$ [1]

As in the last example V values are plotted on a log scale on the vertical axis, whilst the t values are plotted on the horizontal axis on an ordinary linear scale (Fig. 1.12).

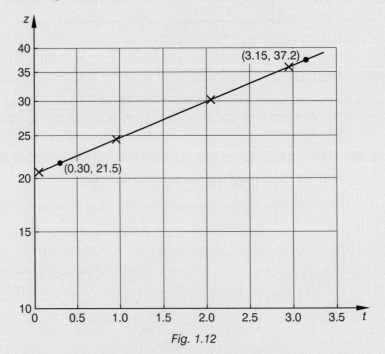

Fig. 1.12

Point $(3.15, 37.2)$ lies on the line, and substituting in equation [1],

$$\lg 37.2 = (0.4343)b(3.15) + \lg a \qquad [2]$$

Point $(0.30, 21.5)$ lies on the line, and substituting in equation [1],

$$\lg 21.5 = (0.4343)b(0.30) + \lg a \qquad [3]$$

Subtracting equation [3] from equation [2],

$$\lg 37.2 - \lg 21.5 = (0.4343)b(3.15 - 0.30)$$

$$\therefore \qquad b = \frac{\lg(37.2/21.5)}{(0.4343)(2.85)}$$

Thus $\qquad b = 0.192$

Substituting in equation [3]

$$\lg 21.5 = (0.4343)(0.192)(0.30) + \lg a$$

$$\therefore \qquad \lg a = 1.3324 - 0.0250$$

$$\therefore \qquad a = 20.3$$

Exercise 1.2

1) Using log–log graph paper show that the following set of values for x and y follows a law of the type $y = ax^n$. From the graph determine the values of a and n.

x	4	16	25	64	144	296
y	6	12	15	24	36	52

2) The following results were obtained in an experiment to find the relationship between the luminosity I of a metal filament lamp and the voltage V:

V	60	80	100	120	140
I	11	20	89	186	319

Allowing for the fact that an error was made in one of the readings show that the law between I and V is of the form $I = aV^n$ and find the probable correct value of the reading. Find the value of n.

3) Two quantities t and m are plotted on log–log graph paper, t being plotted vertically and m being plotted horizontally. The result is a straight line and from the graph it is found that:

when $\qquad\qquad m = 8, \qquad t = 6.8$

and when $\qquad\quad m = 20, \qquad t = 26.9$

Find the law connecting t and m.

4) The intensity of radiation R from certain radioactive materials at a particular time t is thought to follow the law $R = kt^n$. In an experiment to test this the following values were obtained:

R	58	43.5	26.5	14.5	10
t	1.5	2	3	5	7

Show that the assumption was correct and evaluate k and n.

5) The values given in the following table are thought to obey a law of the type $y = ab^{-x}$. Check this statement and find the values of the constants a and b.

x	0.1	0.2	0.4	0.6	1.0	1.5	2.0
y	175	158	60	32	6.4	1.28	0.213

6) The force F on the tight side of a driving belt was measured for different values of the angle of lap θ and the following results were obtained:

F	7.4	11.0	17.5	24.0	36.0
θ rad	$\pi/4$	$\pi/2$	$3\pi/4$	π	$5\pi/4$

Construct a graph to show that these values conform approximately to an equation of the form $F = ke^{\mu\theta}$.

Hence find the constants μ and k.

7) A capacitor and resistor are connected in series. The current i amperes after time t seconds is thought to be given by the equation $i = Ie^{-t/T}$ where I amperes is the initial charging current and T seconds is the time constant. Using the following values verify the relationship and find the values of the constants I and T:

i amperes	0.0156	0.0121	0.009 45	0.007 36	0.005 73
t seconds	0.05	0.10	0.15	0.20	0.25

8) For a constant pressure process on a certain gas the formula connecting the absolute temperature T and the specific entropy s is of the form $T = ke^{cs}$ where e is the logarithmic base and k and c are constants. When $T = 460$, $s = 1.000$, and when $T = 600$, $s = 1.089$. Find constants k and c to three significant figures.

9) The instantaneous e.m.f. v induced in a coil after a time t is given by $v = Ve^{-t/T}$, where V and T are constants. Find the values of V and T given the following values:

v	95	80	65	40	25
t	0.000 13	0.000 56	0.001 08	0.002 29	0.003 47

Graphical Solution of Equations

2

Solving two simultaneous equations – solving a linear and a quadratic equation – finding the roots of a quadratic equation – finding the roots of a cubic equation.

INTRODUCTION

In common with most algebraic functions, simultaneous equations can be displayed as graphs. This enables us to find solutions – generally indicated by the point(s) where the lines or curves intersect. Quadratic equations, which occur from analysis of engineering situations fairly often, have a distinctive shape to their curves – again solutions may be found graphically as an alternative to the algebraic approach.

GRAPHICAL SOLUTION OF TWO SIMULTANEOUS LINEAR EQUATIONS

Since the solutions we require have to satisfy both the given equations, they will be given by the values of the coordinates of the point where the graphs of the equations intersect.

EXAMPLE 2.1

The use of Kirchhoff's laws in an electrical circuit network has resulted in two equations containing currents i_1 and i_2:

$$i_1 - 2i_2 = 2 \tag{1}$$

$$3i_1 + i_2 = 20 \tag{2}$$

Find the values of these currents using a simultaneous graphical solution.

Equation [1] may be written as: $\qquad i_1 = 2i_2 + 2$

and equation [2] may be written as: $\qquad i_1 = -\dfrac{1}{3}i_2 + \dfrac{20}{3}$

Here we have i_1 instead of the usual y, and i_2 replacing x.

i_2			-3	0	3
$i_1 = 2i_2 + 2$			-4	2	8
$i_1 = -\dfrac{1}{3}i_2 + \dfrac{20}{3}$			7.7	6.7	5.7

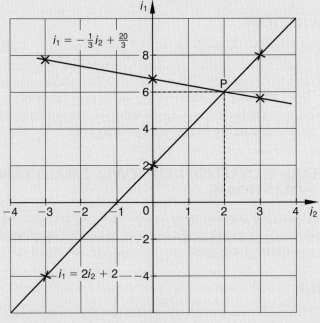

Fig. 2.1

Solving an equation means finding a pair of '(x, y)' values which satisfy the equation. Such a pair of values are the co-ordinates of any point which lies on the graph of the equation. In Fig. 2.1 such a point is P where the two lines intersect. Since it lies on both lines its co-ordinates $(2, 6)$ will satisfy both line equations, and it is, therefore, the required solution.

Hence the two currents are $i_1 = 6$ and $i_2 = 2$.

Exercise 2.1

Solve the following equations:

1) $3x + 2y = 7$ **2)** $4x - 3y = 1$
      $x + y = 3$ $x + 3y = 19$

3) The sum of the ages of two installations is 46 months. The modern version is 10 months younger than the original. Calculate their present ages.

4) A company's annual net profit of £8800 is divided amongst the two partners in the ratio $x : 2y$. If the first shareholder receives £2000 more than the other, find the values of x and y, and hence the respective shares of the profit.

5) The resistance R ohms of a wire at a temperature of t °C is given by the formula $R = R_0(1 + \alpha t)$ where R_0 is the resistance at 0 °C and α is a constant. The resistance is 35 ohms at a temperature of 80 °C and 42.5 ohms at a temperature of 140 °C.   Find R_0 and α. Hence find the resistance when the temperature is 50 °C.

6) A penalty clause states that a contractor will forfeit a certain sum of money for each day that he is late in completing a contract (i.e. the contractor gets paid the value of the original contract less any sum forfeit). If he is 6 days late he receives £5000 and if he is 14 days late he receives £3000. Find the amount of the daily forfeit and determine the value of the original contract.

7) The total cost of equipping two laboratories, A and B, is £30 000. If laboratory B costs £2000 more than laboratory A, find the cost of the equipment for each of them.

8) For a factory winch it is found that the effort E newtons and the load W newtons are connected by the equation $E = aW + b$. An effort of 90 N lifts a load of 100 N whilst an effort of 130 N lifts a load of 200 N. Find the values of a and b and hence determine the effort required to lift a load of 300 N.

GRAPHICAL SOLUTION OF SIMULTANEOUS LINEAR AND QUADRATIC EQUATIONS

Since the solutions we require have to satisfy both the given equations they will be given by the values of x and y where the graphs of the equations intersect.

EXAMPLE 2.2

Solve simultaneously the equations:

$$y = x^2 + 3x - 4$$

and

$$y = 2x + 4$$

We must first draw up tables of values, and will use the range $x = -4$ to $x = +4$:

x		-4	-2	0	2	4
x^2		16	4	0	4	16
$+3x$		-12	-6	0	6	12
-4		-4	-4	-4	-4	-4
$y = x^2 + 3x - 4$		0	-6	-4	6	24
$2x$		-8	-4	0	4	8
$+4$		4	4	4	4	4
$y = 2x + 4$		-4	0	4	8	12

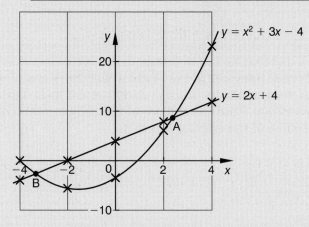

Fig. 2.2

The two graphs are shown plotted on the same axes in Fig. 2.2 and they intersect at the points A and B. Values of the x and y co-ordinates at these points will give the solutions of the given equations. We must be careful not to try to read the values too accurately, as the graphs have been plotted using only five values of x. In this case, even values to the first place of decimals cannot be guaranteed.

If more accurate answers are required then we must plot the portions of the graph containing points A and B using more values of x and also much larger scales.

Hence the required solutions are

x	2.4	-3.4
y	8.7	-2.7

Exercise 2.2

Solve simultaneously:

1) $y = x^2 - 2x - 2$
$x - y + 2 = 0$

2) $y = x^2 - x + 5$
$y = 2x + 5$

3) $y = 5x^2 + x - 3$
$y = 5x - 2$

4) $y = 2x^2 - 2.3x + 1$
$y = 3x - 0.25$

5) A rectangular plot of land has a perimeter of 280 m. The length of a diagonal drawn corner to corner is 100 m. If x and y are the length and width of the plot respectively show that:

$$x + y = 140 \qquad [1]$$

and $$x^2 + y^2 = 10\,000 \qquad [2]$$

Hence find the dimensions of the plot.

GRAPHICAL SOLUTION OF QUADRATIC EQUATIONS

An equation in which the highest power of the unknown is two, and containing no higher powers, is called a quadratic equation. It is also known as an equation of the second degree.
Thus $x^2 - 9 = 0$, $2.5x^2 - 3.1x + 2 = 0$ and $2x^2 - 5x = 0$ are all examples of quadratic equations.

Suppose we plot the graph of $y = x^2 - 9$. It will cut the x-axis when $y = 0$ which is also where $x^2 - 9 = 0$. Thus the x values where the graph cuts the x-axis satisfy the equation $x^2 - 9 = 0$ and these values are therefore roots of the equation.

The examples which follow illustrate this method.

EXAMPLE 2.3

Plot the graph of $y = 3x^2 + 10x - 8$ between $x = -6$ and $x = +4$. Hence solve the equation $3x^2 + 10x - 8 = 0$.

A table can be drawn up as follows giving values of y for the chosen values of x.

x	-6	-5	-4	-3	-2	-1	0	1	2	3	4
$3x^2$	108	75	48	27	12	3	0	3	12	27	48
$10x$	-60	-50	-40	-30	-20	-10	0	10	20	30	40
-8	-8	-8	-8	-8	-8	-8	-8	-8	-8	-8	-8
y	40	17	0	-11	-16	-15	-8	5	24	49	80

The graph of $y = 3x^2 + 10x - 8$ is shown plotted in Fig. 2.3.

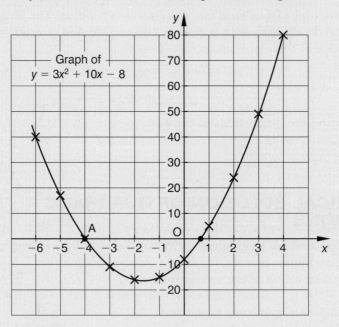

Fig. 2.3

To solve the equation $3x^2 + 10x - 8 = 0$ we have to find the value of x when $y = 0$, that is, the value of x where the graph cuts the x-axis.

These are the points A and B in Fig. 2.3.

Hence the solutions of $3x^2 + 10x - 8 = 0$ are

$$x = -4 \quad \text{and} \quad x = 0.7$$

The accuracy of the results obtained by this method will depend on the scales chosen. The value $x = -4$ is exact as this value was taken when drawing up the table of values, and gave $y = 0$. The value $x = 0.7$ is as accurate as may be read from the scale chosen for the x-axis.

MORE ACCURATE RESULTS

It appears that solution by graph only gives 'fairly rough' answers. It is possible to obtain more accuracy by plotting the portion of the graph in the vicinity of the root to a larger scale. This method is shown in the example which follows.

EXAMPLE 2.4

Find the roots of the equation $x^2 - 1.3 = 0$ by drawing a suitable graph.

Using the same method as in Example 2.3 we need to plot a graph of $y = x^2 - 1.3$ and find where it cuts the x-axis.

We have not been given a range of values of x between which the curve should be plotted and so we must make our own choice.

A good method is to first try a range from $x = -4$ to $x = +4$. If only five values of y are calculated for values of x of -4, -2, 0, $+1$ and $+2$, we shall not have wasted

x	-4	-2	0	2	4
x^2	16	4	0	4	16
-1.3	-1.3	-1.3	-1.3	-1.3	-1.3
y	14.7	2.7	-1.3	2.7	14.7

much time if these values are not required – in any case we shall learn from this trial and be able to make a better choice at the next attempt.

The table of values is shown, and the graph of these values is shown below in Fig. 2.4.

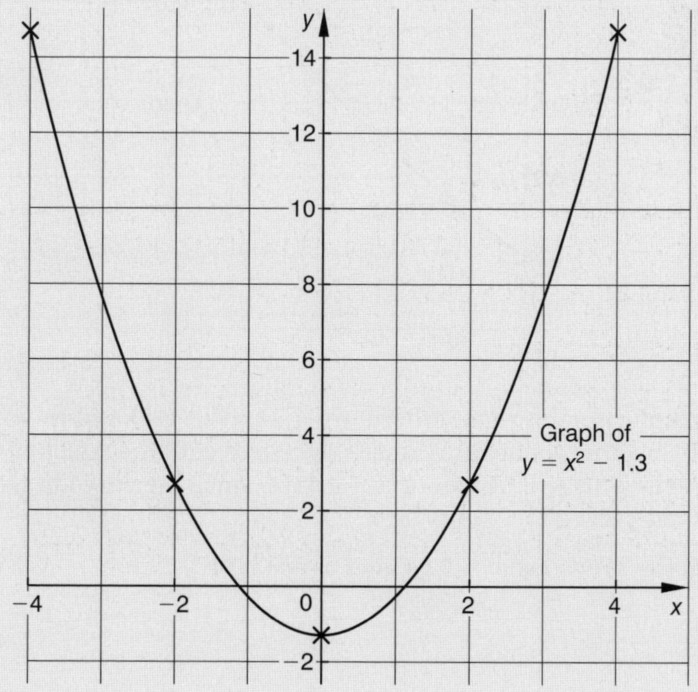

Fig. 2.4

The approximate values of x where the curve cuts the x-axis are -1 and $+1$ (Fig. 2.4). For more accurate results we must plot the portion of the curve where it cuts the x-axis to a larger scale. We can see, however, both from the table of values and the graph, that the graph is symmetrical about the y-axis – so we need only plot one half. We will choose the portion to the right of the y-axis and draw up a table of values from $x = 0.7$ to $x = 1.3$:

x	0.7	0.8	0.9	1.0	1.1	1.2	1.3
x^2 -1.3	0.49 -1.3	0.64 -1.3	0.81 -1.3	1.00 -1.3	1.21 -1.3	1.44 -1.3	1.69 -1.3
y	-0.81	-0.66	-0.49	-0.30	-0.09	0.14	0.39

These values are shown plotted in Fig. 2.5.

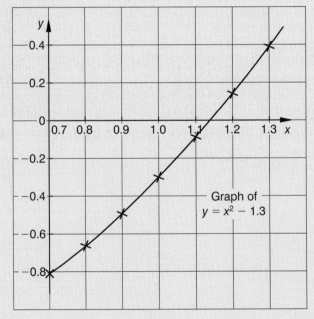

Fig. 2.5

The graph cuts the x-axis where $x = 1.14$ (Fig. 2.5) and we must not forget the other value of x where the curve cuts the x-axis to the left of the y-axis. This will be where $x = -1.14$ since the curve is symmetrical about the y-axis.

Hence the solutions $x = 1.14$ and $x = -1.14$

EXAMPLE 2.5

Solve the equation $x^2 - 4x + 4 = 0$.

We shall plot the graph of $y = x^2 - 4x + 4$ and find where it cuts the x-axis.

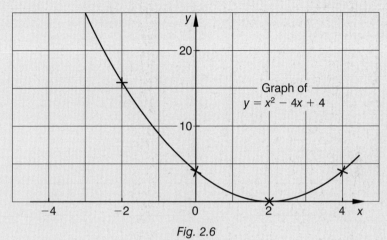

Fig. 2.6

The graph is shown plotted in Fig. 2.6. In this case the curve does not actually cut the x-axis but touches it at the point where $x = 2$. Another way of looking at it is to say that the curve 'cuts' the x-axis at two points which lie on top of each other. The two points coincide and they are said to be coincident points. The roots are called repeated roots.

EXAMPLE 2.6

Solve the equation $x^2 + x + 3 = 0$.

We shall plot the graph of $y = x^2 + x + 3$ and find where it cuts the x-axis.

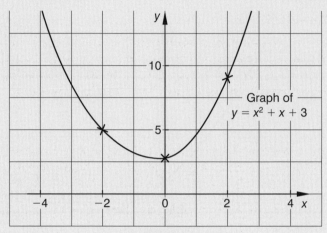

Fig. 2.7

The graph is shown plotted in Fig. 2.7. We can see that the curve does not cut the x-axis at all. This means there are no roots – in theory there are roots but they are complex or imaginary and have no arithmetical value.

Exercise 2.3

By plotting suitable graphs solve the following equations:

1) $x^2 - 7x + 12 = 0$ (plot between $x = 0$ and $x = 6$)

2) $x^2 + 16 = 8x$ (plot between $x = 1$ and $x = 7$)

3) $x^2 - 9 = 0$ (plot between $x = -4$ and $x = 4$)

4) $x^2 + 2x - 15 = 0$ **5)** $3x^2 - 23x + 14 = 0$

6) $2x^2 + 13x + 15 = 0$ **7)** $x^2 - 2x - 1 = 0$

8) $3x^2 - 7x + 1 = 0$ **9)** $9x^2 - 5 = 0$

CUBIC EQUATIONS

An equation in which the highest power of the unknown is three, and containing no higher powers of the unknown, is called a cubic equation. It is also known as an equation of the *third degree*.

Thus $\quad x^3 + x^2 + 7x - 10 = 0, \quad x^3 + 2x^2 + 1 = 0 \quad$ and $\quad x^3 - 37 = 0$ are all examples of cubic equations.

The algebraic method of solving cubic equations is difficult and we shall solve cubic equations by a graphical method similar to that used for solving quadratic equations. This, in fact, is the usual way technicians solve cubic equations when they arise from practical problems.

EXAMPLE 2.7

Plot the graph of $\ y = x^3 - 1.5x^2 - 8.5x + 4.5\ $ from $x = -4$ to $x = +4$ at 1 unit intervals, and use the graph to solve the cubic equation $x^3 - 1.5x^2 - 8.5x + 4.5 = 0$.

A table can be drawn up as follows giving values of y for the chosen values of x:

x	-4	-3	-2	-1	0	1	2	3	4
x^3	-64	-27	-8	-1	0	1	8	27	64
$-1.5x^2$	-24	-13.5	-6	-1.5	0	-1.5	-6	-13.5	-24
$-8.5x$	34	25.5	17	8.5	0	-8.5	-17	-25.5	-34
$+4.5$	4.5	4.5	4.5	4.5	4.5	4.5	4.5	4.5	4.5
y	-49.5	-10.5	7.5	10.5	4.5	-4.5	-10.5	-7.5	10.5

The graph is shown plotted in Fig. 2.8.

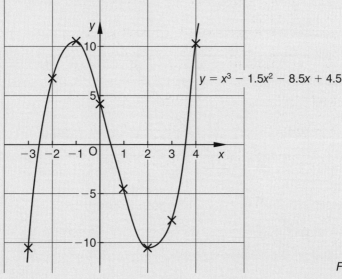

$y = x^3 - 1.5x^2 - 8.5x + 4.5$

Fig. 2.8

To solve the cubic equation we have to find the values of x when $y = 0$, that is, the value of x where the graph cuts the x-axis. Hence the solutions $x = -2.5$, $x = 0.5$ and $x = 3.5$.

We must remember that these values are only approximate, since the values of the first decimal place cannot be guaranteed. As in previous examples, if more accurate answers are required then we must plot the portions of the graph containing the points of intersection using more values of x and also much larger scales.

EXAMPLE 2.8

A domed roof is in the form of a cap of a sphere. Its base radius is 10 m. If the height of the dome is h metres and the volume of air space under the dome is $1525\,\text{m}^3$, it can be shown that

$$h^3 + 300h - 2912 = 0$$

Plot the graph of $y = h^3 + 300h - 2912$ for values of h between 4 and 12 and hence find the value of h.

A table is drawn up giving values of y corresponding to the chosen values of h:

h	4	5	6	7	8	9	10	11	12
h^3	64	125	216	343	512	729	1000	1331	1728
$300h$	1200	1500	1800	2100	2400	2700	3000	3300	3600
-2912	-2912	-2912	-2912	-2912	-2912	-2912	-2912	-2912	-2912
y	-1648	-1287	-896	-469	0	517	1088	1719	2416

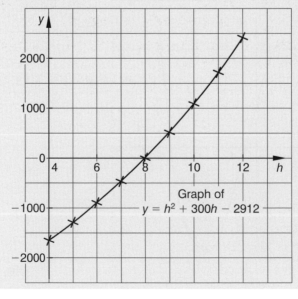

Graph of $y = h^2 + 300h - 2912$

Fig. 2.9

The graph is drawn in Fig. 2.9 and the solution of the equation $h^3 + 300h - 2912 = 0$ is the value of h where the curve cuts the horizontal axis. Hence $h = 8$ is the solution. Note that this solution is exact since the table shows that when $h = 8$, $y = 0$.

EXAMPLE 2.9

Plot the graph of $y = 5x^3 - 9x^2 + 3x + 1$ from $x = -0.4$ to $x = +1.4$ at intervals of 0.2 units, and hence find the values of the roots of the equation $5x^3 - 9x^2 + 3x + 1 = 0$.

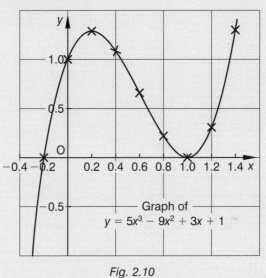

Fig. 2.10

The graph is shown plotted in Fig. 2.10. We can see that the curve cuts the x-axis where $x = -0.2$ and touches the x-axis where $x = 1$.

As in Example 2.5 the point where $x = 1$ represents two coincident points and gives rise to two repeated roots.

Hence the solutions of the equation $5x^3 - 9x^2 + 3x + 1 = 0$ are

$$x = -0.2 \quad \text{and} \quad x = 1$$

EXAMPLE 2.10

Plot the graph of $y = x^3 - 1$ for x values from -1.0 to $+1.5$ at half unit intervals. Hence find the roots of $x^3 - 1 = 0$.

The graph is shown plotted in Fig. 2.11 and it can be seen that the curve only cuts the x-axis at one point which gives the value of the only real root. The other two solutions are complex or imaginary and have no arithmetical meaning.

The only real solution of $x^3 - 1 = 0$ is, therefore, $x = 1$.

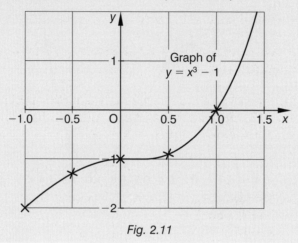

Fig. 2.11

Exercise 2.4

1) Plot the graph of $y = x^3 - 4x^2 - x + 4$ from $x = -2$ to $x = 5$ at one unit intervals. Hence solve the equation $x^3 - 4x^2 - x + 4 = 0$

2) Plot the graph of $y = 3x^3 - 3.4x^2 - 6.4x + 2.4$ using values of x from -2 to $+3$ at half unit intervals. Hence find the roots of the cubic equation $3x^3 - 3.4x^2 - 6.4x + 2.4 = 0$

3) Plot the graph of $y = x^3 + x^2 + x - 3$ from $x = -3$ to $x = +3$ at one unit intervals. Hence show that the cubic equation $x^3 + x^2 + x - 3 = 0$ has only one real root, and find its value.

4) Plot the graph of $y = x^3 - x^2 - x + 1$ taking values of x from -2 to $+2$ at half unit intervals. Hence find the roots of the equation $x^3 - x^2 - x + 1 = 1$

5) Plot the graph of $y = 4x^3 - 15x^2 + 7x + 6$ from $x = -1$ to $x = 3.5$ at intervals of one half unit. Use this graph to solve the cubic equation $4x^3 - 15x^2 + 7x + 6 = 0$

6) Plot the graph of $y = x^3 - x^2 - x - 2$ for values of x from -4 to $+4$ at one unit intervals, and hence find the roots of the equation $x^3 - x^2 - x - 2 = 0$

7) Find the roots of the cubic equation $3x^3 + 4x^2 - 12x + 5 = 0$ by plotting a suitable graph taking values of x from -3.5 to $+1.5$ at half unit intervals.

8) By plotting the graph of $y = x^3 - x^2 - 8x + 12$ find the roots of the equation $x^3 - x^2 - 8x + 12 = 0$

9) Find by graphical means the roots of $7x^3 - 6x^2 - 18x + 4 = 0$.

10) Show that the equation $2x^3 + 3x^2 + 2x + 1 = 0$ has only one real root, and find its value.

11) A spherical vessel has a radius of 8 m. It contains liquid to a height of h metres (Fig. 2.12). When the volume of the liquid in the vessel is 1.056 m^3 the following equation applies:

$$\frac{3168}{\pi} = h^2(24 - h)$$

By plotting a suitable equation find the value of h
(h lies between 3 and 8).

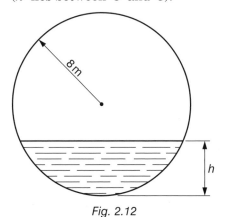

Fig. 2.12

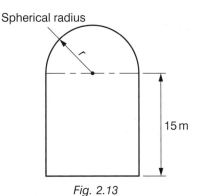

Fig. 2.13

12) A pressure vessel (Fig. 2.13) has a capacity of 594 m^3. The equation from which r may be found is

$$2r^3 + 45r^2 - 567 = 0$$

By drawing a suitable graph find the value of r
(r lies between 0 m and 5 m).

13) A domed roof is in the form of a segment of a sphere. If the volume of air space under the dome is $497\,\text{m}^3$ and the radius of the sphere is $8\,\text{m}$, then

$$h^3 - 24h^2 + 475 \;=\; 0$$

Plot a suitable graph to find h, the height of the dome, given that its value is between $4\,\text{m}$ and $6\,\text{m}$.

14) A cement silo is in the form of a frustum of a cone with a height equal to the larger radius of the frustum. The following equation then applies:

$$R^3 + 2R^2 + 4R - 57 \;=\; 0$$

Find the value of R, the height of the silo, by plotting a suitable graph.

Matrices and Determinants 3

Matrix notation – the sum and difference of two 2×2 matrices – the product of two 2×2 matrices which is, in general, non-commutative – the unit matrix – determinant notation – evaluate a 2×2 determinant – solve simultaneous linear equations with two unknowns – the inverse of a 2×2 matrix – use matrices to solves two simultaneous linear equations with two unknowns.

INTRODUCTION

The object of this chapter is to introduce you to new notation and the rules used in manipulating matrices and determinants.

In mathematics both a matrix and a determinant comprise an arrangement of data, usually numbers and letters, in rows and columns. A matrix array is enclosed by large curved brackets whilst a determinant uses two vertical straight lines.

They are extremely useful tools especially as they lend themselves well to work on computers. The only application here will be to solve two linear simultaneous equations.

ELEMENT

Each number or symbol in a matrix is called an **element** of the matrix.

ORDER

The **dimension** or **order** of a matrix is stated by the number of rows followed by the number of columns in the rectangular array,

e.g.

Matrix	$\begin{pmatrix} 1 & 2 \\ 3 & 4 \end{pmatrix}$	$\begin{pmatrix} a & 2 & -3 \\ 4 & b & x \end{pmatrix}$	$\begin{pmatrix} \sin\theta & 1 \\ \cos\theta & 2 \\ \tan\theta & 3 \end{pmatrix}$	(6)
Order	2×2	2×3	3×2	1×1

EQUALITY

If two matrices are equal, then they must be of the same order and their corresponding elements must be equal.

Thus if $\begin{pmatrix} 2 & 3 & x \\ a & 5 & -2 \end{pmatrix} = \begin{pmatrix} 2 & 3 & 4 \\ -1 & 5 & -2 \end{pmatrix}$ then $x = 4$ and $a = -1$.

ADDITION AND SUBTRACTION

Two matrices may be added or subtracted only if they are of the **same order**. We say the matrices are **conformable** for addition (or subtraction) and we add (or subtract) by combining corresponding elements.

EXAMPLE 3.1

If $\mathbf{A} = \begin{pmatrix} 3 & 4 \\ 5 & 6 \end{pmatrix}$ and $\mathbf{B} = \begin{pmatrix} 0 & 6 \\ 5 & 2 \end{pmatrix}$ determine: a) $\mathbf{C} = \mathbf{A} + \mathbf{B}$

b) $\mathbf{D} = \mathbf{A} - \mathbf{B}$

a) $\mathbf{C} = \begin{pmatrix} 3 & 4 \\ 5 & 6 \end{pmatrix} + \begin{pmatrix} 0 & 6 \\ 5 & 2 \end{pmatrix} = \begin{pmatrix} 3+0 & 4+6 \\ 5+5 & 6+2 \end{pmatrix} = \begin{pmatrix} 3 & 10 \\ 10 & 8 \end{pmatrix}$

b) $\mathbf{D} = \begin{pmatrix} 3 & 4 \\ 5 & 6 \end{pmatrix} - \begin{pmatrix} 0 & 6 \\ 5 & 2 \end{pmatrix} = \begin{pmatrix} 3-0 & 4-6 \\ 5-5 & 6-2 \end{pmatrix} = \begin{pmatrix} 3 & -2 \\ 0 & 4 \end{pmatrix}$

ZERO OR NULL MATRIX

A **zero** or **null** matrix, denoted by **O**, is one in which all the elements are zero. It may be of any order.

Thus $\begin{pmatrix} 0 & 0 \\ 0 & 0 \end{pmatrix}$ is a zero matrix of order 2. It behaves like zero in the real number system.

IDENTITY OR UNIT MATRIX

The **identity** matrix can be of any suitable order with all the main diagonal elements 1 and the remaining elements 0. It is denoted by **I** and behaves like unity in the real number system.

Thus $\begin{pmatrix} 1 & 0 \\ 0 & 1 \end{pmatrix}$ is a unit matrix of order 2.

TRANSPOSE

The **transpose** of a matrix **A** is written as **A′** or **A**^{T}. When the row of a matrix is interchanged with its corresponding column, that is row 1 becomes column 1, and row 2 becomes column 2 and so on, then the matrix is transposed.

Thus if $\mathbf{A} = \begin{pmatrix} 1 & 2 & -3 \\ 4 & 7 & 0 \end{pmatrix}$ then $\mathbf{A}' = \begin{pmatrix} 1 & 4 \\ 2 & 7 \\ -3 & 0 \end{pmatrix}$.

Exercise 3.1

1) State the order of each of the following matrices:

a) $\begin{pmatrix} 1 & 2 \\ 3 & 4 \end{pmatrix}$
b) $\begin{pmatrix} 5 \\ -6 \end{pmatrix}$
c) $\begin{pmatrix} a & b & 4 \\ 2 & 3 & 5 \\ x & -6 & 0 \end{pmatrix}$

d) $\begin{pmatrix} 1 & -2 & -3 & -4 \\ 6 & 2 & 0 & -1 \end{pmatrix}$

2) How many elements are there in:

a) a 3×3 matrix $\qquad$ **b)** a 2×2 matrix

c) a square matrix of order n?

3) Write down the transpose of each matrix in question **1**.

4) Combine the following matrices:

a) $\begin{pmatrix} 2 & 1 \\ 3 & 2 \end{pmatrix} + \begin{pmatrix} -2 & -1 \\ 6 & 0 \end{pmatrix}$ $\qquad$ **b)** $\begin{pmatrix} 2 & 1 \\ 3 & 2 \end{pmatrix} - \begin{pmatrix} -2 & -1 \\ 6 & 0 \end{pmatrix}$

c) $\begin{pmatrix} \frac{1}{2} & 1 \\ \frac{1}{3} & \frac{1}{5} \end{pmatrix} + \begin{pmatrix} \frac{1}{3} & -\frac{1}{2} \\ \frac{1}{2} & \frac{4}{5} \end{pmatrix}$

5) Find a, b and c if $\qquad (a \quad b \quad c) - (-3 \quad 4 \quad 1) = (-5 \quad 1 \quad 0)$

6) Complete $\begin{pmatrix} \frac{1}{2} & \frac{1}{4} \\ \frac{1}{5} & \frac{1}{6} \end{pmatrix} - \begin{pmatrix} \frac{1}{6} & \frac{1}{5} \\ \frac{1}{6} & \frac{1}{9} \end{pmatrix}$

7) Solve the equation $\qquad \mathbf{X} - \begin{pmatrix} 1 & 3 \\ 5 & -2 \end{pmatrix} = \begin{pmatrix} 4 & 5 \\ 7 & 0 \end{pmatrix}$

where $\mathbf{X}$ is a 2×2 matrix.

8) If $\qquad \begin{pmatrix} 4 \\ 5 \end{pmatrix} + \begin{pmatrix} x \\ y \end{pmatrix} = \begin{pmatrix} 4 \\ 10 \end{pmatrix}, \qquad$ determine $\begin{pmatrix} x \\ y \end{pmatrix}$

MULTIPLICATION OF A MATRIX BY A REAL NUMBER

A matrix may be multiplied by a number in the following way:

$$4\begin{pmatrix} 2 & 3 \\ 7 & -1 \end{pmatrix} = \begin{pmatrix} 4 \times 2 & 4 \times 3 \\ 4 \times 7 & 4 \times (-1) \end{pmatrix} = \begin{pmatrix} 8 & 12 \\ 28 & -4 \end{pmatrix}$$

Conversely the common factor of each element in a matrix may be written outside the matrix. Thus $\begin{pmatrix} 9 & 3 \\ 42 & 15 \end{pmatrix} = 3\begin{pmatrix} 3 & 1 \\ 14 & 5 \end{pmatrix}$

MATRIX MULTIPLICATION

Two matrices can only be multiplied together if the number of columns in the first matrix is equal to the number of rows in the second matrix. We say that the matrices are **conformable** for multiplication. The method for multiplying together a pair of 2×2 matrices is as follows:

$$\begin{pmatrix} a & b \\ c & d \end{pmatrix} \times \begin{pmatrix} e & f \\ g & h \end{pmatrix} = \begin{pmatrix} ae + bg & af + bh \\ ce + dg & cf + dh \end{pmatrix}$$

EXAMPLE 3.2

a) $\begin{pmatrix} 2 & 3 \\ 4 & 5 \end{pmatrix} \times \begin{pmatrix} 7 & 1 \\ 0 & 6 \end{pmatrix} = \begin{pmatrix} (2 \times 7) + (3 \times 0) & (2 \times 1) + (3 \times 6) \\ (4 \times 7) + (5 \times 0) & (4 \times 1) + (5 \times 6) \end{pmatrix}$

$$= \begin{pmatrix} 14 & 20 \\ 28 & 34 \end{pmatrix}$$

b) $\begin{pmatrix} 3 & 4 \\ 2 & 5 \end{pmatrix} \times \begin{pmatrix} 6 \\ 7 \end{pmatrix} = \begin{pmatrix} (3 \times 6) + (4 \times 7) \\ (2 \times 6) + (5 \times 7) \end{pmatrix} = \begin{pmatrix} 46 \\ 47 \end{pmatrix}$

c) $\begin{pmatrix} 3 \\ 2 \end{pmatrix} \times \begin{pmatrix} 4 & 6 \\ 5 & 7 \end{pmatrix}$ This is not possible since the matrices are not conformable.

EXAMPLE 3.3

Form the products **AB** and **BA** given that $\mathbf{A} = \begin{pmatrix} 1 & 2 \\ 3 & 4 \end{pmatrix}$ and $\mathbf{B} = \begin{pmatrix} 5 & 6 \\ 7 & 8 \end{pmatrix}$ and hence show that $\mathbf{AB} \neq \mathbf{BA}$.

$\mathbf{AB} = \begin{pmatrix} 1 & 2 \\ 3 & 4 \end{pmatrix} \begin{pmatrix} 5 & 6 \\ 7 & 8 \end{pmatrix} = \begin{pmatrix} (1 \times 5) + (2 \times 7) & (1 \times 6) + (2 \times 8) \\ (3 \times 5) + (4 \times 7) & (3 \times 6) + (4 \times 8) \end{pmatrix}$

$$= \begin{pmatrix} 19 & 22 \\ 43 & 50 \end{pmatrix}$$

$\mathbf{BA} = \begin{pmatrix} 5 & 6 \\ 7 & 8 \end{pmatrix} \begin{pmatrix} 1 & 2 \\ 3 & 4 \end{pmatrix} = \begin{pmatrix} (5 \times 1) + (6 \times 3) & (5 \times 2) + (6 \times 4) \\ (7 \times 1) + (8 \times 3) & (7 \times 2) + (8 \times 4) \end{pmatrix}$

$$= \begin{pmatrix} 23 & 34 \\ 31 & 46 \end{pmatrix}$$

As we see the results are different and, in general, *matrix multiplication is non-commutative*, i.e. $\mathbf{AB} \neq \mathbf{BA}$.

Exercise 3.2

1) If $\mathbf{A} = \begin{pmatrix} 3 & 0 \\ -2 & 1 \end{pmatrix}$ and $\mathbf{B} = \begin{pmatrix} -4 & 1 \\ 3 & -2 \end{pmatrix}$ determine:

    a) $2\mathbf{A}$ **b)** $3\mathbf{B}$ **c)** $2\mathbf{A} + 3\mathbf{B}$ **d)** $2\mathbf{A} - 3\mathbf{B}$

2) Calculate the following products:

    a) $\begin{pmatrix} 3 & 1 \\ 2 & 0 \end{pmatrix}\begin{pmatrix} 4 & -1 \\ 2 & 3 \end{pmatrix}$ **b)** $\begin{pmatrix} 2 & 1 \\ 3 & 1 \end{pmatrix}\begin{pmatrix} 1 & 0 \\ 0 & 1 \end{pmatrix}$

    c) $\begin{pmatrix} 2 & 1 \\ 4 & 2 \end{pmatrix}\begin{pmatrix} 2 & 3 \\ 1 & 5 \end{pmatrix}$ **d)** $\begin{pmatrix} 1 & 0 \\ 0 & 1 \end{pmatrix}\begin{pmatrix} a & b \\ c & d \end{pmatrix}$

    e) $\begin{pmatrix} k & 0 \\ 0 & k \end{pmatrix}\begin{pmatrix} a & b \\ c & d \end{pmatrix}$

3) If $\mathbf{A} = \begin{pmatrix} 1 & 2 \\ 3 & 4 \end{pmatrix}$ and $\mathbf{B} = \begin{pmatrix} 2 & -1 \\ 1 & 3 \end{pmatrix}$ calculate:

    a) $\mathbf{A}^2$ (that is $\mathbf{A} \times \mathbf{A}$)   **b)** $\mathbf{B}^2$ **c)** $2\mathbf{AB}$

    d) $\mathbf{A}^2 + \mathbf{B}^2 + 2\mathbf{AB}$ **e)** $(\mathbf{A} + \mathbf{B})^2$

DETERMINANT OF A SQUARE MATRIX OF ORDER 2

If matrix $\mathbf{A} = \begin{pmatrix} a & b \\ c & d \end{pmatrix}$ then its **determinant** is denoted by $|\mathbf{A}|$ or $\det \mathbf{A}$ and the result is a **number** given by

$$|\mathbf{A}| = \begin{vmatrix} a & b \\ c & d \end{vmatrix} = ad - bc$$

EXAMPLE 3.4

Evaluate $|\mathbf{A}|$ if $\mathbf{A} = \begin{pmatrix} 1 & -2 \\ 3 & 4 \end{pmatrix}$

$$|\mathbf{A}| = \begin{vmatrix} 1 & -2 \\ 3 & 4 \end{vmatrix} = 1 \times 4 - (-2) \times 3 = 10$$

SOLUTION OF SIMULTANEOUS LINEAR EQUATIONS USING DETERMINANTS

To solve simultaneous linear equations with two unknowns using determinants, the following procedure is used.

1) Write out the two equations in order:
$$a_1x + b_1y = c_1$$
$$a_2x + b_2y = c_2$$

2) Calculate $\Delta = \begin{vmatrix} a_1 & b_1 \\ a_2 & b_2 \end{vmatrix}$

3) Then $x = \dfrac{\begin{vmatrix} c_1 & b_1 \\ c_2 & b_2 \end{vmatrix}}{\Delta}$ and $y = \dfrac{\begin{vmatrix} a_1 & c_1 \\ a_2 & c_2 \end{vmatrix}}{\Delta}$

EXAMPLE 3.5

By using determinants, solve the simultaneous equations

$$3x + 4y = 22$$
$$2x + 5y = 24$$

Now $\Delta = \begin{vmatrix} 3 & 4 \\ 2 & 5 \end{vmatrix} = (3 \times 5) - (4 \times 2) = 7$

Thus $x = \dfrac{\begin{vmatrix} 22 & 4 \\ 24 & 5 \end{vmatrix}}{7} = \dfrac{(22 \times 5) - (4 \times 24)}{7} = \dfrac{14}{7} = 2$

And $y = \dfrac{\begin{vmatrix} 3 & 22 \\ 2 & 24 \end{vmatrix}}{7} = \dfrac{(3 \times 24) - (22 \times 2)}{7} = \dfrac{28}{7} = 4$

Exercise 3.3

1) Evaluate the following determinants:

a) $\begin{vmatrix} 5 & 2 \\ 3 & 6 \end{vmatrix}$ **b)** $\begin{vmatrix} 7 & 4 \\ 5 & 2 \end{vmatrix}$ **c)** $\begin{vmatrix} 6 & 8 \\ 2 & 5 \end{vmatrix}$

2) Solve the following simultaneous equations by using determinants:

a) $3x + 4y = 11$
  $x + 7y = 15$

b) $5x + 3y = 29$
  $4x + 7y = 37$

c) $4x - 6y = -2.5$
  $7x - 5y = -0.25$

THE INVERSE OF A SQUARE MATRIX OF ORDER 2

Instead of dividing a number by 5 we can multiply by $\frac{1}{5}$ and obtain the same result.

Thus $\frac{1}{5}$ is the multiplicative inverse of 5. That is $5 \times \frac{1}{5} = 1$.

In matrix algebra we never divide by a matrix but multiply instead by the inverse. The inverse of matrix $\mathbf{A}$ is denoted by $\mathbf{A}^{-1}$ and is such that

$$\mathbf{AA}^{-1} = \begin{pmatrix} 1 & 0 \\ 0 & 1 \end{pmatrix} = \mathbf{I}, \text{ the identity matrix}$$

To find the inverse $\mathbf{A}^{-1}$ of the square matrix $\mathbf{A} = \begin{pmatrix} a & b \\ c & d \end{pmatrix}$ we use the expression:

$$\mathbf{A}^{-1} = \frac{1}{|\mathbf{A}|} \begin{pmatrix} d & -b \\ -c & a \end{pmatrix} = \frac{1}{ad - bc} \begin{pmatrix} d & -b \\ -c & a \end{pmatrix}$$

EXAMPLE 3.6

Determine the inverse of $\mathbf{A} = \begin{pmatrix} 1 & -2 \\ 3 & 4 \end{pmatrix}$ and verify the result.

Now
$$|\mathbf{A}| = \begin{vmatrix} 1 & -2 \\ 3 & 4 \end{vmatrix} = (1 \times 4) - (3 \times -2) = 10$$

Hence
$$\mathbf{A}^{-1} = \tfrac{1}{10} \begin{pmatrix} 4 & 2 \\ -3 & 1 \end{pmatrix} = \begin{pmatrix} 0.4 & 0.2 \\ -0.3 & 0.1 \end{pmatrix}$$

To verify the result we have

$$\mathbf{AA}^{-1} = \begin{pmatrix} 1 & -2 \\ 3 & 4 \end{pmatrix} \begin{pmatrix} 0.4 & 0.2 \\ -0.3 & 0.1 \end{pmatrix} = \begin{pmatrix} 1 & 0 \\ 0 & 1 \end{pmatrix} = \mathbf{I}$$

SINGULAR MATRIX

A matrix which does not have an inverse is called a **singular matrix**. This happens when $|\mathbf{A}| = 0$.

For example, since $\begin{vmatrix} 3 & 6 \\ 1 & 2 \end{vmatrix} = (3 \times 2) - (6 \times 1) = 0$

then $\begin{pmatrix} 3 & 6 \\ 1 & 2 \end{pmatrix}$ is a singular matrix.

Exercise 3.4

Decide whether each of the matrices in questions **1–9** has an inverse. If the inverse exists, find it.

1) $\begin{pmatrix} 2 & 5 \\ 1 & 4 \end{pmatrix}$

2) $\begin{pmatrix} 2 & 5 \\ 1 & 3 \end{pmatrix}$

3) $\begin{pmatrix} 3 & 2 \\ 1 & 2 \end{pmatrix}$

4) $\begin{pmatrix} 4 & 10 \\ 2 & 5 \end{pmatrix}$

5) $\begin{pmatrix} 224 & 24 \\ 24 & 4 \end{pmatrix}$

6) $\begin{pmatrix} a & -b \\ -a & b \end{pmatrix}$

7) $\begin{pmatrix} 2 & 3 \\ -1 & 1 \end{pmatrix}$

8) $\begin{pmatrix} 2 & -3 \\ 1 & 5 \end{pmatrix}$

9) $\begin{pmatrix} 1 & 1 \\ 0 & 1 \end{pmatrix}$

10) Given that $\mathbf{A} = \begin{pmatrix} 1 & 0 \\ 3 & 2 \end{pmatrix}$ and $\mathbf{B} = \begin{pmatrix} 3 & 5 \\ 1 & 2 \end{pmatrix}$, calculate:

    a) $\mathbf{A}^{-1}$ **b)** $\mathbf{B}^{-1}$ **c)** $\mathbf{B}^{-1}\mathbf{A}^{-1}$ **d)** $\mathbf{AB}$

    e) $(\mathbf{AB})^{-1}$ **f)** Compare the answers to **c)** and **e)**.

SYSTEMS OF LINEAR EQUATIONS

Given the system of equations $\left.\begin{array}{r} 5x + y = 7 \\ 3x - 4y = 18 \end{array}\right\}$

we can rewrite it in the form $\begin{pmatrix} 5x + y \\ 3x - 4y \end{pmatrix} = \begin{pmatrix} 7 \\ 18 \end{pmatrix}$

or $\begin{pmatrix} 5 & 1 \\ 3 & -4 \end{pmatrix}\begin{pmatrix} x \\ y \end{pmatrix} = \begin{pmatrix} 7 \\ 18 \end{pmatrix}$

That is $\begin{pmatrix} \text{Matrix of} \\ \text{coefficients} \end{pmatrix}\begin{pmatrix} \text{Matrix of} \\ \text{variables} \end{pmatrix} = \begin{pmatrix} \text{Matrix of} \\ \text{constants} \end{pmatrix}$

Denote the matrix of coefficients by $\mathbf{C}$ and its inverse by $\mathbf{C}^{-1}$.

Then $\quad |\mathbf{C}| = \begin{vmatrix} 5 & 1 \\ 3 & -4 \end{vmatrix} = 5 \times (-4) - 1 \times 3 = -23$

and $\quad \mathbf{C}^{-1} = \dfrac{1}{-23}\begin{pmatrix} -4 & -1 \\ -3 & 5 \end{pmatrix} = \dfrac{1}{23}\begin{pmatrix} 4 & 1 \\ 3 & -5 \end{pmatrix}$

Now
$$\mathbf{C}\begin{pmatrix} x \\ y \end{pmatrix} = \begin{pmatrix} 7 \\ 18 \end{pmatrix}$$

and multiplying both sides by $\mathbf{C}^{-1}$ gives

$$\mathbf{C}^{-1}\mathbf{C}\begin{pmatrix} x \\ y \end{pmatrix} = \mathbf{C}^{-1}\begin{pmatrix} 7 \\ 18 \end{pmatrix}$$

$\therefore$
$$\mathbf{I}\begin{pmatrix} x \\ y \end{pmatrix} = \mathbf{C}^{-1}\begin{pmatrix} 7 \\ 18 \end{pmatrix}$$

or
$$\begin{pmatrix} 1 & 0 \\ 0 & 1 \end{pmatrix} \times \begin{pmatrix} x \\ y \end{pmatrix} = \frac{1}{23}\begin{pmatrix} 4 & 1 \\ 3 & -5 \end{pmatrix} \times \begin{pmatrix} 7 \\ 18 \end{pmatrix}$$

$\therefore$
$$\begin{pmatrix} 1 \times x & 0 \times y \\ 0 \times x & 1 \times y \end{pmatrix} = \frac{1}{23}\begin{pmatrix} 4 \times 7 + & 1 \times 18 \\ 3 \times 7 + (-5) \times 18 \end{pmatrix}$$

$\therefore$
$$\begin{pmatrix} x \\ y \end{pmatrix} = \frac{1}{23}\begin{pmatrix} 46 \\ -69 \end{pmatrix}$$

$\therefore$
$$\begin{pmatrix} x \\ y \end{pmatrix} = \begin{pmatrix} 2 \\ -3 \end{pmatrix}$$

Thus comparing the matrices shows that $x = 2$ and $y = -3$.

We would not normally perform multiplication by the unit matrix.

We did so here to illustrate that when a matrix, here $\begin{pmatrix} x \\ y \end{pmatrix}$, is multiplied by the unit matrix then it is unaltered.

This confirms that the unit matrix performs as unity (the number one) in normal arithmetic.

Exercise 3.5

Use matrix methods to solve each of the following systems of equations:

1) $\begin{aligned} x + y &= 1 \\ 3x + 2y &= 8 \end{aligned}$

2) $\begin{aligned} x + y &= 6 \\ 3x - 2y &= -7 \end{aligned}$

3) $\begin{aligned} 5x - 2y &= 17 \\ 2x + 3y &= 3 \end{aligned}$

4) $\begin{aligned} 3x - 2y &= 12 \\ 4x + y &= 5 \end{aligned}$

5) $\begin{aligned} 3x + 2y &= 6 \\ 4x - y &= 5 \end{aligned}$

6) $\begin{aligned} 3x - 4y &= 26 \\ 5x + 6y &= -20 \end{aligned}$

Geometric and Arithmetic Progressions

4

Arithmetic progression (A.P.) – any term – cutting and spindle speeds – sum of an A.P. – simple interest
Geometric progression (G.P.) – any term – comparison of spindle speeds in A.P. and G.P. – sum of a G.P. – sum to infinity – convergence – compound interest.

INTRODUCTION

A sequence of numbers, which are connected by a definite mathematical law, is called a progression or series. The following are examples:

1, 3, 5, 7, ... (each term is obtained by adding 2 to the preceding one),
1, 2, 4, 8, ... (each term is twice the preceding one),
1, 4, 9, 16, 25, ... (the terms are the squares of successive integers).

Geometric and arithmetical progressions are two of the more common number series.

Applications include machine tool cutting speeds. Also, applicable to us both personally and in business, is simple and compound interest.

THE ARITHMETIC PROGRESSION

A series in which each term is obtained by adding or subtracting a constant quantity is called an **arithmetic progression** (or just **A.P.**). Thus the terms of the progression increase or decrease by a fixed amount which is called the **common difference**.

In the series   3, 6, 9, 12, 15, ...,

the difference between each term and the preceding one is 3, i.e. the common difference is 3. The series is therefore an A.P.

The series   10, 8, 6, 4, 2, 0, -2, -4, -6, ...

is an A.P. since each term is 2 less than the preceding one, i.e. the common difference is -2.

General Expression for a Series in A.P.

Let the first number in the series be a and the common difference be d. The series can then be written as

$$a, \quad a + d, \quad a + 2d, \quad a + 3d, \ ...$$

and hence any term of the series, say the nth term, is

$$\boxed{a + (n - 1)d}$$

EXAMPLE 4.1

Find the 9th term of the series 1, 5, 9, 13, ...

The 1st term of the series is 1, and so $a = 1$

The common difference is 4, and so $d = 4$

Hence the 9th term is

$$a + (9 - 1)d \ = \ 1 + (8 \times 4) \ = \ 33$$

EXAMPLE 4.2

The first term of a series in A.P. is 7 and the fifth term is 19. Find the eleventh term.

The 1st term $a = 7$ and   the 5th term $= a + (5 - 1)d = a + 4d$

Hence $7 + 4d \ = \ 19$

$$d \ = \ 3$$

11th term $= a + (11 - 1)d \ = \ 7 + 10 \times 3 \ = \ 37$

Cutting Speeds

The **cutting speed** for a centre lathe is the distance moved by a point on the surface of the work (such as point A in Fig. 4.1) in one minute. This is the speed at which the work moves past the tool point. It is also called the **surface speed** of the work. Cutting speeds are always quoted in metres per minute.

Let

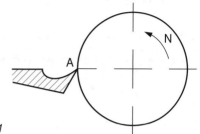

D(mm)   = diameter of work,
N(rev/min) = spindle speed of lathe
S(m/min)   = cutting speed

Fig. 4.1

Now the distance moved by point A over the surface in one rev is

$$\pi D \text{ mm} \;=\; \frac{\pi D}{1000}\,\text{m}$$

Then the distance moved by point A over the surface in N revs is $\dfrac{\pi DN}{1000}$ metres.

Hence   cutting speed $\boxed{S \;=\; \dfrac{\pi DN}{1000}}$

This formula can be transposed to give

$$\boxed{N \;=\; \frac{1000S}{\pi D}} \quad \text{and} \quad \boxed{D \;=\; \frac{1000S}{\pi N}}$$

When milling or drilling the same formulae for cutting speeds etc. may be used, but then D would represent the diameter of the cutter or drill.

Speeds of Spindles and Feed Shafts; Approximations to Whole Numbers

These may be obtained exactly using the formulae developed. However, the cutting speeds, although recommended by the manufacturers of materials and tools as the best values, are only approximate. Therefore in the calculations of spindle speeds etc. the figures obtained are rounded off to the nearest whole numbers. This is also the practice on the gearboxes of most machine tools where the whole numbers of teeth on gears also limit accuracy.

In the examples which follow in this chapter the exact values will be rounded off to the nearest whole numbers.

Arithmetic Progressions applied to Spindle Speeds

In designing a machine tool the highest and lowest spindle speeds are usually determined by the extremes in work sizes which the machine will be required to handle. For example, a lathe may be designed to take a range of work varying from 25 mm to 375 mm diameter.

If we allow for a cutting speed of 20 m/min then the highest spindle speed, suitable for 25 mm diameter to the nearest whole number, is:

$$N = \frac{1000 \times 20}{\pi \times 25} = 255 \,\text{rev/min}$$

The lowest spindle speed, suitable for 375 mm diameter work, is

$$N = \frac{1000 \times 20}{\pi \times 375} = 17 \,\text{rev/min}$$

Suppose that the lathe has to have eight spindle speeds. Six intermediate spindle speeds must be chosen so that we have a series. One way of doing this is to choose the speeds so that they form a series in A.P.

the first term $a = 17$ and the eighth term $\qquad a + 7d = 255$

giving $\qquad\qquad\qquad\qquad\qquad\qquad\qquad\qquad d = 34$

Hence all the speeds can be found, the second speed being

$$a + d = 17 + 34 = 51$$

and so on, giving the eight speeds as follows:

$$17, 51, 85, 119, 153, 187, 221 \text{ and } 255 \,\text{rev/min}$$

Sum of a Series in Arithmetic Progression

Consider the typical A.P. series already met, supposing it to consist of n terms of which the last term is l.

$$
\begin{array}{ll}
\text{The 1st term is} & a \\
\text{the 2nd term is} & a + d \\
\text{the 3rd term is} & a + 2d, \text{ etc.}
\end{array}
$$

and the last term is $l = a + (n - 1)d$

The sum of the series is $\quad S = a + (a + d) + (a + 2d) + \ldots\ldots + l$

and backwards $\quad\quad\quad\quad S = l + (l - d) + (l - 2d) + \ldots\ldots + a$

so adding, term by term, $2S = (a + 1) + (a + 1) + \ldots\ldots\ldots + (a + l)$

since the d terms cancel out. There are n terms so there are n lots of $(a + l)$ giving:

$$2S = n(a + l)$$

or

$$\boxed{S = \frac{n}{2}(a + l) \quad\quad \text{where} \quad\quad l = a + (n - 1)d}$$

EXAMPLE 4.3

Find the sum of the series 1.5, 2.6, 3.7, 4.8, ..., if there are altogether 12 terms.

The last term $\quad\quad\quad\quad\quad l = a + (n - 1)d$

In this case $a = 1.5$, $n = 12$ and $d = 1.1$ (the difference between any two adjacent terms), and substituting these values in the expression for l we get

$$l = 1.5 + (12 - 1)1.1 = 1.5 + 12.1 = 13.6$$

But the sum $\quad\quad S = \frac{n}{2}(a + l) \quad\quad$ and on substituting our values,

$$S = \frac{12}{2}(1.5 + 13.6) = 6 \times 15.1 = 90.6$$

Exercise 4.1

1) Find the 8th and 17th terms of the series 12, 15, 18, 21, ...

2) Which term of the series -40, -38, -36, ... is equal to 4?

3) The 3rd term of an A.P. is -8 and the 16th is 57. Find
   a) the 10th term,
   b) the sum of the first 14 terms.

4) Insert six arithmetic means between 3.30 and 6.45

5) Three numbers are in A.P. The product of the 1st term and the last term is -140. The sum of three times the 2nd term and twice the 1st term is -14. Find the numbers.

6) How many terms in the series 5.6, 4.1, 2.6, ... must be taken so that their sum is -3.6?

7) In boring a well 300 m deep, the cost of boring the first metre is £200 000 whilst each subsequent metre costs £250. What is the cost of boring the entire well?

8) Water fills a tank at a rate of 150 litres during the first hour, 350 litres during the second hour, 550 litres during the third hour and so on. Find the number of hours necessary to fill a rectangular tank 16 m × 7 m × 7 m.

9) A body falling freely falls 5 m in the first second of its motion, 15 m in the second second, 25 m during the third second and so on.
   a) How far does it fall in 12 s?
   b) How far does it fall in the 14th second?
   c) How long will it take to fall 3000 m?

10) A lathe has to have five spindle speeds arranged in A.P. It is to cater for work ranging from 25 mm to 200 mm in diameter. Allowing for a cutting speed of 25 m/min, find the spindle speeds.

THE GEOMETRIC PROGRESSION

A series in which each term is obtained from the preceding term by multiplying or dividing by a constant quantity is called a **geometric progression** or simply a **G.P.** The constant quantity is called the **common ratio** of the series.

In the series 1, 2, 4, 8, 16, ..., each successive term is formed by multiplying the preceding one by 2. The series is therefore in G.P., and the common ratio is 2.

The series 6, 2, 2/3, 2/9, ... is a G.P. since each successive term is formed from the preceding by multiplying by $\frac{1}{3}$, and this is also the common ratio.

A General Expression for a Series in G.P.

Let the first term be a and the common ratio r. The series can be represented by a, ar, ar^2, ar^3, ...

We see that: the 1st term is a,

the 2nd term is ar (the index of r is 1)

the 3rd term is ar^2 (the index of r is 2)

the 4th term is ar^3 (the index of r is 3) etc.

The index of r is always one less than the number of the term in the series.

Hence   the nth term is $\boxed{ar^{n-1}}$

EXAMPLE 4.4

Find the seventh term of the series 2, 6, 18, ...

The 1st term is 2 and the common ratio is 3, that is, $a = 2$ and $r = 3$.
Hence the 7th term is $ar^6 = 2 \times 3^6 = 1458$

EXAMPLE 4.5

The 1st term of a series in G.P. is 19, and the 6th is 27. Find the 10th term.

$$\text{1st term} \qquad a = 19$$
$$\text{and the 6th term} = ar^5 = 27$$
$$\therefore \qquad 19r^5 = 27$$
$$\therefore \qquad r = \sqrt[5]{\frac{27}{19}} = 1.073$$
$$\text{Hence the 10th term} = ar^9 = 19 \times 1.073^9 = 35.8$$

EXAMPLE 4.6

Insert four geometric means between 1.8 and 11.2

This is equivalent to putting four terms between 1.8 and 11.2 so that the six numbers form a series in G.P. Thus 1.8, ?, ?, ?, ?, 11.2 must be a G.P.

The 1st term is 1.8 $\qquad a = 1.8$

The 6th term is 11.2 $\qquad ar^5 = 11.2$

that is, $\qquad 1.8r^5 = 11.2$

$\therefore \qquad r = \sqrt[5]{\dfrac{11.2}{1.8}} = 1.441$

Hence the 2nd term $\qquad ar = 1.8 \times 1.441 = 2.59$

the 3rd term, $\qquad ar^2 = 1.8 \times 1.441^2 = 3.74$

the 4th term, $\qquad ar^3 = 1.8 \times 1.441^3 = 5.39$

and the 5th term, $\qquad ar^4 = 1.8 \times 1.441^4 = 7.76$

Hence the required geometric means are 2.59, 3.74, 5.39 and 7.76

Geometric Progressions Applied to Spindle Speeds

Previously we have considered spindle speeds in A.P. They may also be arranged in G.P. and, as will be seen later, this is a better way of arranging them.

EXAMPLE 4.7

A drill is to have seven speeds arranged in G.P. It is to drill a range of holes from 3 mm to 12 mm in diameter at a cutting speed of 15 m/min. Find the seven spindle speeds.

Since $N = \dfrac{1000S}{\pi D}$, $\qquad$ the 1st speed $= \dfrac{1000 \times 15}{\pi \times 12} = 398$ rev/min

$\qquad$ and the 7th speed $= \dfrac{1000 \times 15}{\pi \times 3} = 1592$ rev/min

Since all the speeds are to be in G.P., $\quad a = 398$

$\therefore \qquad 398r^6 = 1592$

$\therefore \qquad r = \sqrt[6]{\dfrac{1592}{398}} = 1.26$

Hence the 2nd speed is $ar = 398 \times 1.26 = 501 \,\text{rev/min}$
and the 3rd speed is $ar^2 = 398 \times 1.26^2 = 632 \,\text{rev/min}$

which are rounded off to the nearest whole numbers giving the required speeds as

$$398, 501, 632, 796, 1003, 1264 \text{ and } 1592 \,\text{rev/min}$$

COMPARISON OF SPINDLE SPEEDS IN A.P. AND IN G.P.

In this section we shall show that spindle speeds in G.P. are to be preferred to those in A.P. This will be done by considering the following example:

EXAMPLE 4.8

A lathe tool has to accommodate work between 25 mm and 300 mm in diameter. Six spindle speeds are required and the cutting speed is to be 25 m/min. Find the six speeds, (a) if they are in A.P., (b) if they are in G.P. In both cases find the work diameter appropriate to each spindle speed.

(a) *Speeds in A.P.*

$$1\text{st speed} = \frac{1000 \times 25}{\pi \times 300} = 27 \,\text{rev/min} = a$$

$$6\text{th speed} = \frac{1000 \times 25}{\pi \times 25} = 318 \,\text{rev/min} = a + 5d$$

$\therefore$ $\qquad\qquad 27 + 5d = 318$

$\therefore$ $\qquad\qquad\qquad d = 58.2$

$\therefore$ $\qquad 2\text{nd speed} = a + d = 85 \,\text{rev/min}$

$\qquad\text{and } 3\text{rd speed} = a + 2d = 143 \,\text{rev/min}$

Hence arranged in A.P. the spindle speeds are

$$27, 85, 143, 202, 260 \text{ and } 318 \,\text{rev/min}$$

The bar diameters appropriate to each speed can be calculated by the formula

$$D = \frac{1000 \times S}{\pi \times N}$$

These are, in order, 295, 94, 56, 39, 31 and 25 mm.

(b) *Speeds in G.P.*

$$\text{1st speed } a = 27$$
$$\text{and 6th speed} = ar^5 = 318$$
$$\therefore \qquad 27r^5 = 318$$
$$\therefore \qquad r = \sqrt[5]{\frac{318}{27}} = 1.638$$
$$\therefore \qquad \text{2nd speed} = ar = 44$$
$$\text{and 3rd speed} = ar^2 = 72$$

Hence, arranged in G.P., the spindle speeds are

$$27, 44, 72, 119, 194 \text{ and } 318 \text{ rev/min}$$

The corresponding bar diameters can be calculated as before and are, in order, 295, 181, 111, 67, 41 and 25 mm.

For convenience the figures obtained for the speeds arranged in both A.P. and G.P. are tabulated below:

Speed number	A.P.		G.P.	
	Speed	Bar diam.	Speed	Bar diam.
1	27	295	27	295
2	85	94	44	181
3	143	56	72	111
4	202	39	119	67
5	260	31	194	41
6	318	25	318	25

The graph (Fig. 4.2) shows that when the speeds are in A.P. there are too many at the higher end of the speed range and too few at the lower end. If we wish, for instance, to turn a 175 mm diameter bar the correct spindle speed would be 45 rev/min to give a cutting speed of 25 m/min. We can either use speed number 1 (27 rev/min) or speed number 2 (85 rev/min). Speed number 1 is too low and speed number 2 is too high. Spindle speeds in G.P. give a far better distribution. For any diameter we can find a spindle speed which will give a near approximation to the correct cutting speed of 25 m/min. It is for this reason that the spindle speeds are usually arranged in G.P. rather than in A.P.

The graph (Fig. 4.3) shows spindle speed plotted against speed number. It can be seen that the graph is straight for A.P. and curved for G.P. This illustrates the better distribution of spindle speeds when they are arranged in G.P. The G.P. arrangement gives closer speed intervals at the lower end of the range and wider intervals at the higher end.

Fig. 4.2

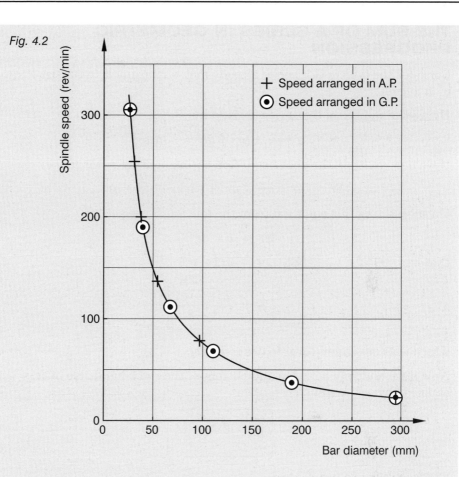

Fig. 4.3

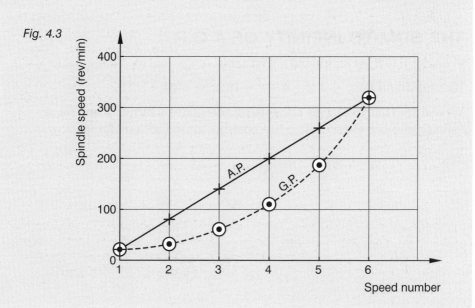

THE SUM OF A SERIES IN GEOMETRIC PROGRESSION

Consider the typical G.P. series already met, supposing it to consist of n terms.

Hence the sum S of the series is given by

$$S = a + ar + ar^2 + ar^3 + \ldots + ar^{n-1}$$

If we now multiply the above equation throughout by r we get

$$Sr = ar + ar^2 + ar^3 + \ldots + ar^{n-1} + ar^n$$

On subtracting this equation from the one for S we get

$$S - Sr = a - ar^n$$

that is,
$$S(1 - r) = a(1 - r^n)$$

$\therefore$
$$S = \frac{a(1 - r^n)}{1 - r}$$

which is a convenient form to use if $r < 1$.

Now if we multiply both top and bottom of the right-hand side of this equation by -1 we have

$$S = \frac{a(r^n - 1)}{r - 1}$$

which is easier to use if $r > 1$.

THE SUM TO INFINITY OF A G.P.

Consider a G.P. series having a first term $a = 1$ and a common ratio $r = 2$.

Its sum would be $\quad 1 + 2 + 4 + 8 + 16 + 32 + 64 + \ldots$

We can see that the terms are getting successively larger. Thus for an infinite number of terms it is *not* possible to find a value for their sum.

However, for a G.P. series having $a = 1$ and $r = \dfrac{1}{3}$

the sum would be $\quad 1 + \dfrac{1}{3} + \dfrac{1}{9} + \dfrac{1}{27} + \dfrac{1}{81} + \dfrac{1}{243} + \dfrac{1}{729} + \ldots$

Here the terms are rapidly getting smaller and a numerical answer to any reasonable degree of accuracy may soon be obtained. Such a series is said to be **convergent**.

In general for a G.P. series having $r < 1$, then it is convergent. If we consider the expression for the sum $\quad S = \dfrac{a(1 - r^n)}{1 - r} \quad$ then for an infinite series (which means that n is infinitely large) the term r^n will be virtually zero. Thus the sum to infinity will be $\quad \dfrac{a}{1 - r}$.

Note. An A.P. series will always be non-convergent, so it is *not* possible to find its sum to infinity.

EXAMPLE 4.9

Find the sum of the series 2, 6, 18, ... which has 6 terms.

Since $r = 3$ (obtained from 6/2 or 18/6, etc.),
that is $r > 1$, we shall use the form $\quad S = \dfrac{a(r^n - 1)}{r - 1}$

In this question $a = 2$ and $n = 6$.
Substituting these values we get

$$S = \frac{2(3^6 - 1)}{3 - 1}$$

$$= \frac{2(729 - 1)}{2}$$

$$= 728$$

EXAMPLE 4.10

In the first week of production 1000 articles are made. If the rise in weekly production is 5% per week, how many weeks are necessary to produce a total of 15 000 articles?

In the 1st week,   production $= 1000$

In the 2nd week,   production $= 1000 + 1000\left(\dfrac{5}{100}\right) = 1000\left(1 + \dfrac{5}{100}\right)$

In the 3rd week.   production $= 1000\left(1 + \dfrac{5}{100}\right)^2$

the weekly production rates form a series:

$$1000, \quad 1000\left(1 + \dfrac{5}{100}\right), \quad 1000\left(1 + \dfrac{5}{100}\right)^2, \quad \text{etc.,}$$

that is, $1000, \quad 1000(1.05), \quad 1000(1.05)^2, \quad \text{etc.,}$

which is a G.P. having $a = 1000$ and $r = 1.05$

Now the sum $S = \dfrac{a(r^n - 1)}{r - 1}$

We have to find the number of weeks n for $S = 15\,000$

$\therefore$ $15\,000 = \dfrac{1000(1.05^n - 1)}{1.05 - 1}$

from which $1.05^n = 1.75$

To solve this equation for n we take logs of both sides.

$$\log 1.05^n = \log 1.75$$

$\therefore$ $n \times \log 1.05 = \log 1.75$

$\therefore$ $n = \dfrac{\log 1.75}{\log 1.05}$

$\therefore$ $n = 11.47 \text{ weeks.}$

COMPOUND INTEREST

Consider a sum of £P, invested at $r\%$ for n years.

The initial value is £P, and after 1 year the value is

$$£\left(P + \frac{Pr}{100}\right) \quad \text{or} \quad £P\left(1 + \frac{r}{100}\right)$$

In the 2nd year the interest will be on

$$£P\left(1 + \frac{r}{100}\right)$$

so after the 2nd year the value is

$$£\left[P\left(1 + \frac{r}{100}\right) + P\left(1 + \frac{r}{100}\right) \times \frac{r}{100}\right]$$

$$= £\left[P\left(1 + \frac{r}{100}\right)\left(1 + \frac{r}{100}\right)\right]$$

$$= £P\left(1 + \frac{r}{100}\right)^2$$

Similarly after the 3rd year the value is

$$£P\left(1 + \frac{r}{100}\right)^3$$

and the amount after n years is obtained from the G.P. whose first term is £P and whose common ratio is $(1 + r/100)$, and equals

$$\boxed{£P\left(1 + \frac{r}{100}\right)^n}$$

EXAMPLE 4.11

Calculate the value of £2500 invested at 5% compound interest after eight years.

Here $P = 2500$, $r = 5$ and $n = 8$, and substituting these values in the formula £$P(1 + r/100)^n$ we find the required value to be

$$£2500\left(1 + \frac{5}{100}\right)^8$$

$$= £2500(1.05)^8$$

$$= £3693.64$$

Exercise 4.2

1) Find the 8th term of the series 1, 1.1, 1.21, ...

2) Find the 8th term of the series which is a G.P. having a 2nd term of -3 and a 5th term of 81.

3) Insert 3 geometric means between $1\frac{1}{8}$ and $\frac{1}{72}$

4) Find the sum of the series 60, 30, 15, ...,
   a) to 6 terms,   **b)** to infinity.

5) The sum of the first 6 terms of a G.P. series is 189 and the common ratio is 2. Find the series.

6) On a certain lathe the cone and back gear give 8 possible spindle speeds. If the least speed is 3 rev/min and the greatest speed is 300 rev/min, find the intermediate spindle speeds, if they are in G.P.

7) A lathe has 6 speeds arranged in G.P.   It is to accommodate work from 12 mm diameter to 150 mm diameter at a cutting speed of 22 m/min. Find the six spindle speeds.

8) A drill has to drill holes between 2 mm and 25 mm in diameter. Eight spindle speeds are required and the cutting speed is to be 20 m/min. Find the eight speeds if
   a) they are in A.P.,   **b)** they are in G.P.

   Illustrate your results by means of a graph, on which the spindle speeds (rev/min) are plotted vertically and bar diameters (mm) are plotted horizontally. Comment on the results, stating which system is preferable and why.

9) A contractor hires out machinery. In the first year of hiring out one piece of equipment the profit is £6000, but this diminishes by 5% on successive years. Show that the annual profits form a G.P. and find the total of all profits for
   a) the first 6 years,   **b)** all possible future years.

10) A firm starts work with 110 employees for the 1st week. The number of employees rises by 6% per week.
   a) How many weeks will it take for 250 persons to be employed?
   b) How many persons will be employed in the 20th week if the present rate of expansion continues?

11) Car production in the first week of a new model was 150. If it continues with a fall per week of 2%.
   a) find the total output in 52 weeks from the commencement of production in the new model,
   b) find how many cars will be produced in the next 52 weeks.

The Binomial Theorem

A binomial expression – the binomial theorem – the binomial coefficient.

INTRODUCTION

A binomial expression is one which has two terms, for example $a + b$. The binomial theorem gives any power of this expression.

We shall use the theorem later in work on statistics and so it is important to have a knowledge of this topic.

BINOMIAL EXPRESSION

A **binomial expression** consists of two terms. Thus
$$1 + x, \quad a + b, \quad 5y - 2, \quad 3x^2 + 7 \quad \text{and} \quad 7a^3 + 3b^2$$
are all binomial expressions.
The **binomial theorem** allows us to expand powers of such expressions.

THE BINOMIAL THEOREM

Now $$(a + b)^0 = 1$$

(since any number to the power 0 is unity),

and also $$(a + b)^1 = a + b$$

Multiplying both sides by $(a + b)$ gives

$$(a + b)^2 = a^2 + 2ab + b^2$$

also $$(a + b)^3 = a^3 + 3a^2b + 3ab^2 + b^3$$

and $$(a + b)^4 = a^4 + 4a^3b + 6a^2b^2 + 4ab^3 + b^4$$

We can now arrange the coefficients of each of the terms of the above expansions in the form known as **Pascal's triangle**.

Binomial expression	Coefficients in the expansion

$$
\begin{array}{c}
(a+b)^0 \\
(a+b)^1 \\
(a+b)^2 \\
(a+b)^3 \\
(a+b)^4 \\
(a+b)^5 \\
(a+b)^6 \\
(a+b)^7
\end{array}
\qquad
\begin{array}{ccccccccccccccc}
 & & & & & & & 1 & & & & & & & \\
 & & & & & & 1 & & 1 & & & & & & \\
 & & & & & 1 & & 2 & & 1 & & & & & \\
 & & & & 1 & & 3 & & 3 & & 1 & & & & \\
 & & & 1 & & 4 & & 6 & & 4 & & 1 & & & \\
 & & 1 & & 5 & & 10 & & 10 & & 5 & & 1 & & \\
 & 1 & & 6 & & 15 & & 20 & & 15 & & 6 & & 1 & \\
1 & & 7 & & 21 & & 35 & & 35 & & 21 & & 7 & & 1
\end{array}
$$

It will be seen that:

a) The number of terms in each expansion is one more than the index. Thus the expansion of $(a+b)^9$ will have 10 terms.

b) The arrangement of the coefficients is symmetrical.

c) The coefficients of the first and last terms are both always unity.

d) Each coefficient in the table is obtained by adding together the two coefficients in the line above that lie on either side of it.

The expansion of $(a+b)^8$ is therefore:

$$a^8 + (1+7)a^7b + (7+21)a^6b^2 + (21+35)a^5b^3 + (35+35)a^4b^4$$
$$+ (35+21)a^3b^5 + (21+7)a^2b^6 + (7+1)ab^7 + b^8$$

$$= a^8 + 8a^7b + 28a^6b^2 + 56a^5b^3 + 70a^4b^4 + 56a^3b^5 + 28a^2b^6 + 8ab^7 + b^8$$

It is inconvenient to use Pascal's triangle when expanding higher powers of $(a+b)$. In such cases the following series is used:

$$(a+b)^n = a^n + na^{n-1}b + \frac{n(n-1)}{2!}a^{n-2}b^2 + \frac{n(n-1)(n-2)}{3!}a^{n-3}b^3$$
$$+ \ldots + b^n$$

This is the **binomial theorem**
and is true for all positive whole numbers n.

The symbol '!' indicates 'factorial' when following a positive whole number.

For example, 4! is pronounced 'factorial four' and means $4 \times 3 \times 2 \times 1$.

Thus $2! = 2 \times 1$ $3! = 3 \times 2 \times 1$
      $4! = 4 \times 3 \times 2 \times 1$ $5! = 5 \times 4 \times 3 \times 2 \times 1$ and so on.

EXAMPLE 5.1

Expand $(3x + 2y)^4$

Comparing $(3x + 2y)^4$ with $(a + b)^4$, we have $3x$ in place of a, and $2y$ in place of b. Substituting in the standard expansion, we get

$$[(3x) + (2y)]^4 = (3x)^4 + 4(3x)^3(2y) + \frac{(4)(4-1)}{2!}(3x)^2(2y)^2$$

$$+ \frac{(4)(4-1)(4-2)}{3!}(3x)(2y)^3 + (2y)^4$$

$$= (3x)^4 + 4(3x)^3 2y + \frac{4 \times 3}{2 \times 1}(3x)^2(2y)^2$$

$$+ \frac{4 \times 3 \times 2}{3 \times 2 \times 1}(3x)(2y)^3 + (2y)^4$$

$$= 81x^4 + 216x^3y + 216x^2y^2 + 96xy^3 + 16y^4$$

EXAMPLE 5.2

Expand $(x - 4y)^{15}$ to four terms

The given binomial expression should be rewritten as $[x + (-4y)]^{15}$, and comparing this with the standard expression for $(a + b)^n$ we have x in place of a, $-4y$ in place of b, and 15 in place of n.

$$\therefore \quad [x + (-4y)]^{15} = x^{15} + 15x^{14}(-4y) + \frac{(15)(15-1)}{2!}x^{13}(-4y)^2$$

$$+ \frac{15(15-1)(15-2)}{3!}x^{12}(-4y)^3 + \ldots$$

$$= x^{15} + 15(-4)x^{14}y + \frac{15 \times 14 \times (-4)^2}{2 \times 1}x^{13}y^2$$

$$+ \frac{15 \times 14 \times 13 \times (-4)^3}{3 \times 2 \times 1}x^{12}y^3 + \ldots$$

$$= x^{15} - 60x^{14}y + 1680x^{13}y^2 - 29\,120x^{12}y^3 + \ldots$$

THE BINOMIAL COEFFICIENT

One of the uses of this topic is for work on probability which you will meet later in this book. It is often necessary to be able to evaluate the coefficient of any term even if the value of n is reasonably large.

We have, for the binomial theorem expansion of $(a + b)^n$ an expression for

> the coefficient of the $(r + 1)$th term: $\dfrac{n!}{(n - r)! \; r!}$

EXAMPLE 5.3

Find the coefficient of the sixth term of the expansion of $(a + b)^8$.

Here we have $n = 8$ and, since $r + 1 = 6$, then $r = 5$.

Thus the coefficient of the 6th term

$$= \frac{8!}{(8 - 5)! \; 5!} = \frac{8!}{3! \times 5!} = \frac{8 \times 7 \times 6 \times 5 \times 4 \times 3 \times 2 \times 1}{3 \times 2 \times 1 \quad \times \quad 5 \times 4 \times 3 \times 2 \times 1} = 56$$

verifying a result found earlier.

Many calculators have a key for 'factorial' so a sequence could be

$\boxed{AC}$ $\boxed{8}$ $\boxed{x!}$ $\boxed{\div}$ $\boxed{3}$ $\boxed{x!}$ $\boxed{\div}$ $\boxed{5}$ $\boxed{x!}$ $\boxed{=}$ giving 56

Beware of a trap – it is easy to use multiply instead of divide before entering $5!$ – decide for yourself why this would be wrong.

Exercise 5.1

Expand

  1) $(1 + z)^5$ **2)** $(p + q)^6$ **3)** $(x - 3y)^4$

  4) $(2p\backslash - q)^5$ **5)** $(2x + y)^7$ **6)** $\left(x + \dfrac{1}{x}\right)^3$

Expand to four terms using the binomial theorem:

  7) $(1 + x)^{12}$ **8)** $(1 - 2x)^{14}$ **9)** $(p + q)^{16}$

10) $(1 + 3y)^{10}$ **11)** $(x^2 - 3y)^9$ **12)** $\left(x^2 + \dfrac{1}{x^2}\right)^{11}$

13) Find the fourth term of the expansion for $(a + b)^7$

14) Find the seventh term of the expansion of $(a + b)^{10}$

15) Find the ninth term of the expansion for $(a + b)^{18}$

Complex Numbers

Define j as $\sqrt{-1}$ – definition of a complex number – the algebraic (Cartesian) form – roots of a quadratic equation – add and subtract numbers in algebraic form – the conjugate of a complex number – multiply and divide numbers in algebraic form – represent as a phasor a complex number on an Argand diagram – the j operator – add and subtract phasors on an Argand diagram – the polar form of complex number – conversion of polar form to algebraic form, and vice versa.

INTRODUCTION

Complex numbers are based on the use of the square root of -1. We know that no two numbers multiplied together can give negative result so we just give $\sqrt{-1}$ a single letter name j and you will see how this helps to further our ideas. One of the main uses is in defining alternating currents and their phase differences with voltage.

A QUADRATIC EQUATION WITH COMPLEX ROOTS

The solution of the quadratic equation $ax^2 + bx + c = 0$ is given by the formula

$$x = \frac{-b \pm \sqrt{b^2 - 4ac}}{2a}$$

When we use this formula most of the quadratic equations we meet, when solving technology problems, are found to have roots which are ordinary positive or negative numbers.

Consider now the equation $x^2 - 4x + 13 = 0$.

Then
$$x = \frac{-(-4) \pm \sqrt{(-4)^2 - 4 \times 1 \times 13}}{2 \times 1}$$

$$= \frac{4 \pm \sqrt{-36}}{2}$$

$$= \frac{4 \pm \sqrt{(-1)(36)}}{2}$$

$$= \frac{4 \pm \sqrt{(-1)} \times \sqrt{(36)}}{2}$$

$$= \frac{4}{2} \pm \sqrt{-1} \times \frac{6}{2}$$

$$= 2 \pm \sqrt{-1} \times 3$$

It is not possible to find the value of the square root of a negative number.

In order to try to find a meaning for roots of this type we represent $\sqrt{-1}$ by the symbol j.

(Books on pure mathematics often use the symbol i, but in technology j is preferred as i is used for the instantaneous value of a current.)

Thus the roots of the above equation become $2 + j3$ and $2 - j3$.

DEFINITIONS

Expressions such as $2 + j3$ are called **complex numbers**. The number 2 is called the **real part** and j3 is called the **imaginary part**.

The general expression for a complex number is $x + jy$, which has a real part equal to x and an imaginary part equal to jy. The form $x + jy$ is said to be the **algebraic form** of a complex number. It may also be called the **Cartesian form** or **rectangular notation**.

POWERS OF j

We have defined j such that

$$j = \sqrt{-1}$$

∴ squaring both sides of the equation gives

$$j^2 = (\sqrt{-1})^2 = -1$$

Hence

$$j^3 = j^2 \times j = (-1) \times j = -j$$

and

$$j^4 = (j^2)^2 = (-1)^2 = 1$$

and

$$j^5 = j^4 \times j = 1 \times j = j$$

and $$j^6 = (j^2)^3 = (-1)^3 = -1$$

and so on.

The most used of the above relationships is $j^2 = -1$.

ADDITION AND SUBTRACTION OF COMPLEX NUMBERS IN ALGEBRAIC FORM

The real and imaginary parts must be treated separately. The real parts may be added and subtracted and also the imaginary parts may be added and subtracted, both obeying the ordinary laws of algebra.

Thus
$$(3 + j2) + (5 + j6) = 3 + j2 + 5 + j6$$
$$= (3 + 5) + j(2 + 6)$$
$$= 8 + j8$$

and
$$(1 - j2) - (-4 + j) = 1 - j2 + 4 - j$$
$$= (1 + 4) - j(2 + 1)$$
$$= 5 - j3$$

EXAMPLE 6.1

If z_1, z_2 and z_3 represent three complex numbers such that $z_1 = 1.6 + j2.3$, $z_2 = 4.3 - j0.6$ and $z_3 = -1.1 - j0.9$ find the complex numbers which represent:

a) $z_1 + z_2 + z_3$,

b) $z_1 - z_2 - z_3$.

a) $z_1 + z_2 + z_3 = (1.6 + j2.3) + (4.3 - j0.6) + (-1.1 - j0.9)$
$$= 1.6 + j2.3 + 4.3 - j0.6 - 1.1 - j0.9$$
$$= (1.6 + 4.3 - 1.1) + j(2.3 - 0.6 - 0.9)$$
$$= 4.8 + j0.8$$

b) $z_1 - z_2 - z_3 = (1.6 + j2.3) - (4.3 - j0.6) - (-1.1 - j0.9)$
$$= 1.6 + j2.3 - 4.3 + j0.6 + 1.1 + j0.9$$
$$= (1.6 - 4.3 + 1.1) + j(2.3 + 0.6 + 0.9)$$
$$= -1.6 + j3.8$$

MULTIPLICATION OF COMPLEX NUMBERS IN ALGEBRAIC FORM

Consider the product of two complex numbers, $(3 + j2)(4 + j)$.

The brackets are treated in exactly the same way as they are in ordinary algebra, such that

$$(a + b)\,(c + d) \;=\; ac + bc + ad + bd$$

Hence
$$
\begin{aligned}
(3 + j2)(4 + j) &= 3 \times 4 + j2 \times 4 + 3 \times j + j2 \times j \\
&= 12 + j8 + j3 + j^2 2 \\
&= 12 + j8 + j3 - 2 \quad \text{since} \quad j^2 = -1 \\
&= (12 - 2) + j(8 + 3) \\
&= 10 + j11
\end{aligned}
$$

EXAMPLE 6.2

Express the product of $2 + j$, $-3 + j2$, and $1 - j$ as a single complex number.

Then
$$
\begin{aligned}
(2 + j)(-3 + j2)(1 - j) &= (2 + j)(-3 + j2 + j3 - j^2 2) \\
&= (2 + j)(-1 + j5) \quad \text{since} \quad j^2 = -1 \\
&= -2 - j + j10 + j^2 5 \\
&= -7 + j9 \quad\quad\quad\quad \text{since} \quad j^2 = -1
\end{aligned}
$$

CONJUGATE COMPLEX NUMBERS

Consider
$$
\begin{aligned}
(x + jy)(x - jy) &= x^2 + jxy - jxy - j^2 y \\
&= x^2 - (-1)y^2 \\
&= x^2 + y^2
\end{aligned}
$$

Hence we have the product of two complex numbers which produces a real number since it does not have a j term. If $x + jy$ represents a complex number then $x - jy$ is known as its **conjugate** (and vice versa). For example, the conjugate of $(3 + j4)$ is $(3 - j4)$ and their product is

$$(3 + j4)(3 - j4) = 9 + j12 - j12 - j^2 16 = 9 - (-1)16$$
$$= 25 \quad \text{which is a real number}$$

DIVISION OF COMPLEX NUMBERS IN ALGEBRAIC FORM

Consider $\dfrac{(4+j5)}{(1-j)}$.

We use the method of rationalising the denominator.

This means removing the j terms from the bottom line of the fraction. If we multiply $(1-j)$ by its conjugate $(1+j)$ the result will be a real number. Hence, in order not to alter the value of the given expression, we multiply both the numerator and the denominator by $(1+j)$.

Thus
$$\frac{(4+j5)}{(1-j)} = \frac{(4+j5)(1+j)}{(1-j)(1+j)}$$

$$= \frac{4+j5+j4+j^2 5}{1-j+j-j^2}$$

$$= \frac{4+j9+(-1)5}{1-(-1)}$$

$$= \frac{-1+j9}{2}$$

$$= -0.5+j4.5$$

EXAMPLE 6.3

The impedance Z of a circuit having a resistance and inductive reactance in series is given by the complex number $Z = 5+j6$
Find the admittance of the circuit if $Y = 1/Z$

Now
$$Y = \frac{1}{Z} = \frac{1}{5+j6}$$

The conjugate of the denominator is $5-j6$ and we therefore multiply both the numerator and denominator by $5-j6$.

Then
$$Y = \frac{(5-j6)}{(5+j6)(5-j6)}$$

$$= \frac{5-j6}{25+j30-j30-j^2 36}$$

$$= \frac{5-j6}{25-(-1)36} = \frac{5-j6}{61}$$

$$= 0.082 - j0.098$$

EXAMPLE 6.4

Two impedances Z_1 and Z_2 are given by the complex numbers $Z_1 = 1 + j5$ and $Z_2 = j7$. Find the equivalent impedance Z if

a) $Z = Z_1 + Z_2$ where Z_1 and Z_2 are in series,

b) $\dfrac{1}{Z} = \dfrac{1}{Z_1} + \dfrac{1}{Z_2}$ when Z_1 and Z_2 are in parallel.

a)
$$
\begin{aligned}
Z = Z_1 + Z_2 &= (1 + j5) + j7 \\
&= 1 + j5 + j7 \\
&= 1 + j12
\end{aligned}
$$

b)
$$
\begin{aligned}
\frac{1}{Z} = \frac{1}{Z_1} + \frac{1}{Z_2} &= \frac{1}{(1 + j5)} + \frac{1}{j7} \\
&= \frac{j7 + (1 + j5)}{(1 + j5)j7} \\
&= \frac{1 + j12}{j7 + j^2 35} \\
&= \frac{1 + j12}{j7 + (-1)35}
\end{aligned}
$$

Thus
$$
\begin{aligned}
Z &= \frac{j7 - 35}{1 + j12} \\
&= \frac{(j7 - 35)(1 - j12)}{(1 + j12)(1 - j12)} \\
&= \frac{j7 - 35 - j^2 84 + j420}{1 + j12 - j12 - j^2 144} \\
&= \frac{j427 - 35 - (-1)84}{1 - (-1)144} \\
&= \frac{49 + j427}{145} \\
&= 0.338 + j2.945
\end{aligned}
$$

Exercise 6.1

1) Add the following complex numbers:

   a) $3 + j5$, $7 + j3$ and $8 + j2$

   b) $2 - j7$, $3 + j8$ and $-5 - j2$

   c) $4 - j2$, $7 + j3$, $-5 - j6$ and $2 - j5$.

2) Subtract the following complex numbers:

   a) $3 + j5$ from $2 + j8$

   b) $7 - j6$ from $3 - j9$

   c) $-3 - j5$ from $7 - j8$.

3) Simplify the following expressions giving the answers in the form $x + jy$:

   a) $(3 + j3)(2 + j5)$ **b)** $(2 - j6)(3 - j7)$

   c) $(4 + j5)^2$ **d)** $(5 + j3)(5 - j3)$

   e) $(-5 - j2)(5 + j2)$ **f)** $(3 - j5)(3 - j3)(1 + j)$

   g) $\dfrac{1}{2 + j5}$ **h)** $\dfrac{2 + j5}{2 - j5}$

   i) $\dfrac{-2 - j3}{5 - j2}$ **j)** $\dfrac{7 + j3}{8 - j3}$

   k) $\dfrac{(1 + j2)(2 - j)}{(1 + j)}$ **l)** $\dfrac{4 + j2}{(2 + j)(1 - j3)}$

4) Find the real and imaginary parts of:

   a) $1 + \dfrac{j}{2}$ **b)** $j3 + \dfrac{2}{j^3}$ **c)** $(j2)^2 + 3(j)^5 - j(j)$

5) Solve the following equations giving the answers in the form $x + jy$:

   a) $x^2 + 2x + 2 = 0$ **b)** $x^2 + 9 = 0$

6) Find the admittance Y of a circuit if $Y = \dfrac{1}{Z}$ where $Z = 1.3 + j0.6$

7) Three impedances Z_1, Z_2 and Z_3 are represented by the complex numbers $Z_1 = 2 + j$, $Z_2 = 1 + j$ and $Z_3 = j2$. Find the equivalent impedance Z if:

   a) $Z = Z_1 + Z_2 + Z_3$ **b)** $\dfrac{1}{Z} = \dfrac{1}{Z_1} + \dfrac{1}{Z_2} + \dfrac{1}{Z_3}$

   c) $Z = \dfrac{1}{\dfrac{1}{Z_1} + \dfrac{1}{Z_2}} + Z_3$

THE ARGAND DIAGRAM

When plotting a graph, Cartesian co-ordinates are generally used to plot the points. Thus the position of the point P (Fig. 6.1) is defined by the co-ordinates (3, 2) meaning that $x = 3$ and $y = 2$.

Complex numbers may be represented in a similar way on the Argand diagram. The real part of the complex number is plotted along the horizontal real-axis whilst the imaginary part is plotted along the vertical imaginary, or j-axis.

However, a complex number is denoted not by a point but as a **phasor**. A phasor is a line in which regard is paid both to its magnitude and to its direction. Hence in Fig. 6.2 the complex number $4 + j3$ is represented by the phasor $\overrightarrow{OQ}$, the end Q of the line being found by plotting 4 units along the real-axis and 3 units along the j-axis.

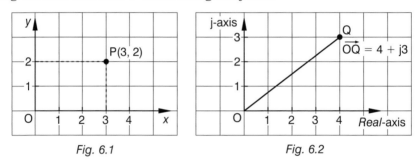

Fig. 6.1   Fig. 6.2

A single letter, the favourite being z, is often used to denote a phasor which represents a complex number. Thus if $z = x + jy$ it is understood that z represents a phasor and not a simple numerical value.

Four typical complex numbers z_1, z_2, z_3 and z_4 are shown on the Argand diagram in Fig. 6.3.

A real number such as 2.7 may be regarded as a complex number with a zero imaginary part, i.e. $2.7 + j0$, and may be represented on the Argand diagram (Fig. 6.4) as the phasor $z = 2.7$ denoted by $\overrightarrow{OA}$ in the diagram.

A number such as j3 is said to be wholly imaginary and may be regarded as a complex number having a zero real part, i.e. $0 + j3$, and may be represented on the Argand diagram (Fig. 6.4) as the phasor $z = j3$ denoted by $\overrightarrow{OB}$ in the diagram.

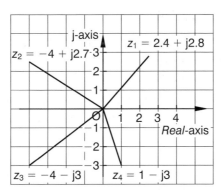

Fig. 6.3

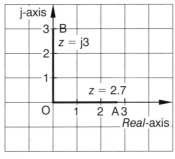

Fig. 6.4

THE j-OPERATOR

Consider the real number 3 shown on the Argand diagram in Fig. 6.5.

It may be denoted by $\overrightarrow{OA}$. (This is a phasor because it has magnitude and direction.)

If we now multiply the real number 3 by j we obtain the complex number $j3$ which may be represented by the phasor $\overrightarrow{OB}$.

It follows that the effect of j on phasor $\overrightarrow{OA}$ is to make it become phasor $\overrightarrow{OB}$,

that is $$\overrightarrow{OB} = j\overrightarrow{OA}$$

Hence j is known as an **operator** (called the **j-operator**) which, when applied to a phasor, alters its direction by $90°$ in an anti-clockwise direction without changing its magnitude.

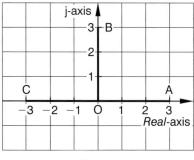

Fig. 6.5

If we now operate on the phasor $\overrightarrow{OB}$ we shall obtain, therefore, phasor $\overrightarrow{OC}$.

In equation form this is

$$\overrightarrow{OC} = j\overrightarrow{OB}$$

but since $\overrightarrow{OB} = j\overrightarrow{OA}$, then

$$\overrightarrow{OC} = j(j\overrightarrow{OA})$$
$$= j^2\overrightarrow{OA}$$
$$= -\overrightarrow{OA} \qquad \text{since } j^2 = -1$$

This is true since it may be seen from the phasor diagram that phasor $\overrightarrow{OC}$ is equal in magnitude, but opposite in direction, to phasor $\overrightarrow{OA}$.

Consider now the effect of the j-operator on the complex number $5 + j3$.

In the equation form this is

$$j(5 + j3) = j5 + j(j3)$$
$$= j5 + j^23$$
$$= j5 + (-1)3$$
$$= -3 + j5$$

If phasor $z_1 = 5 + j3$ and phasor $z_2 = -3 + j5$, it may be seen from the Argand diagram in Fig. 6.6 that their magnitudes are the same but the effect of the operator j on z_1 has been to alter its direction by $90°$ anticlockwise to give phasor z_2.

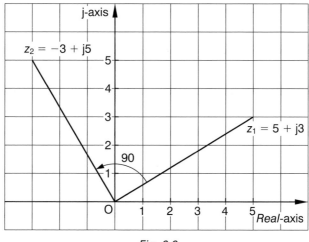

Fig. 6.6

ADDITION OF PHASORS

Consider the addition of the two complex numbers $2 + j3$ and $4 + j2$.

We have
$$(2 + j3) + (4 + j2) = 2 + j3 + 4 + j2$$
$$= (2 + 4) + j(3 + 2)$$
$$= 6 + j5$$

On the Argand diagram shown in Fig. 6.7, the complex number $2 + j3$ is represented by the phasor $\overrightarrow{OA}$, whilst $4 + j2$ is represented by phasor $\overrightarrow{OB}$. The addition of the real parts is performed along the real-axis and the addition of the imaginary parts is carried out on the j-axis.

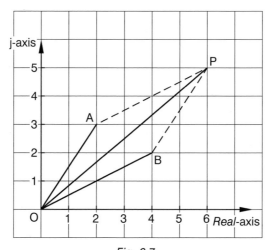

Fig. 6.7

Hence the complex number $6 + j5$ is represented by the phasor $\overrightarrow{OP}$.

It follows that

$$\overrightarrow{OP} = \overrightarrow{OB} + \overrightarrow{OA}$$

Hence the addition of phasors is similar to vector addition used when dealing with forces or velocities.

SUBTRACTION OF PHASORS

Consider the difference of the two complex numbers $4 + j5$ and $1 + j4$.

We have
$$\begin{aligned}(4 + j5) - (1 + j4) &= 4 + j5 - 1 - j4 \\ &= (4 - 1) + j(5 - 4) \\ &= 3 + j\end{aligned}$$

On the Argand diagram shown in Fig. 6.8, the complex number $4 + j5$ is represented by the phasor $\overrightarrow{OC}$, whilst $1 + j4$ is represented by the phasor $\overrightarrow{OD}$. The subtraction of the real parts is performed along the real-axis, and the subtraction of the imaginary parts is carried out along the j-axis. Now let $(4 + j5) - (1 + j4) = 3 + j$ be represented by the phasor $\overrightarrow{OQ}$.

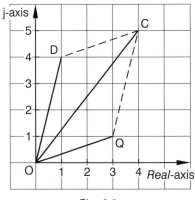

Fig. 6.8

It follows that
$$\overrightarrow{OQ} = \overrightarrow{OC} - \overrightarrow{OD}$$

As for phasor addition, the subtraction of phasors is similar to the subtraction of vectors.

THE POLAR FORM OF A COMPLEX NUMBER

Let z denote the complex number represented by the phasor $\overrightarrow{OP}$ shown in Fig. 6.9. Then from the right-angled triangle PMO we have

$$z = x + jy$$
$$= r\cos\theta + j(r\sin\theta)$$
$$= r(\cos\theta + j\sin\theta)$$

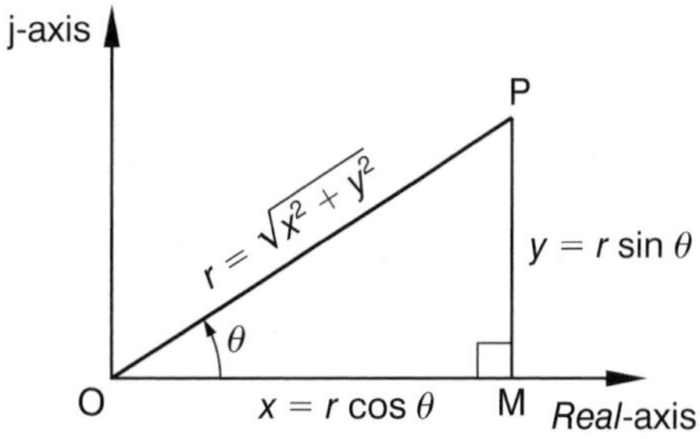

Fig. 6.9

The expression $r(\cos\theta + j\sin\theta)$ is known as the **polar form** of the complex number z. Using conventional notation it may be shown abbreviated as $r\underline{/\theta}$.

r is called the **modulus** of the complex number z and is denoted by $\mod z$ or $|z|$.

Hence, from the diagram, $|z| = r = \sqrt{x^2 + y^2}$, using the theorem of Pythagoras for the right-angled triangle PMO.

It should be noted that the plural of 'modulus' is 'moduli'.

The angle θ is called the **argument** (or amplitude) of the complex number z and is denoted by $\arg z$ (or $\text{amp}\, z$).

Hence $$\arg z = \theta$$

and, from the diagram $$\tan\theta = \frac{y}{x}$$

There are an infinite number of angles whose tangents are the same, and so it is necessary to define which value of θ to state when solving the equation $\tan\theta = \frac{y}{x}$. It is called the **principal value** of the angle and lies between $+180°$ and $-180°$.

We recommend that, when finding the polar form of a complex number, you should sketch it on an Argand diagram. This will help you to avoid a common error of giving an incorrect value of the angle.

EXAMPLE 6.5

Find the modulus and argument of the complex number $3 + j4$ and express the complex number in polar form.

Let $z = 3 + j4$ which is shown in the Argand diagram in Fig. 6.10.

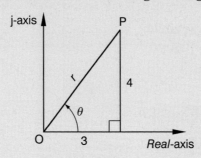

Fig. 6.10

Then $$|z| = r = \sqrt{3^2 + 4^2} = 5$$

and $$\tan \theta = \tfrac{4}{3} = 1.3333$$

∴ $$= 53.13°$$

Hence in polar form

$$z = 5(\cos 53.13° + j \sin 53.13°)$$

or $$z = 5 \angle 53.13°$$

EXAMPLE 6.6

Show the complex number $z = 5 \angle -150°$ on an Argand diagram, and find z in algebraic form.

Now z is represented by phasor $\overrightarrow{OP}$ in Fig. 6.11. It should be noted that since the angle is negative it is measured in a clockwise direction from the real-axis datum.

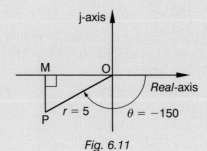

Fig. 6.11

In order to express z in algebraic form we need to find the lengths MO and MP. We use the right-angled triangle PMO in which $\stackrel{\wedge}{\text{POM}} = 180° - 150° = 30°$.

Now $\qquad\qquad$ MO $=$ PO cos $\stackrel{\wedge}{\text{POM}} = 5\cos 30° = 4.33$

and $\qquad\qquad$ MP $=$ PO sin $\stackrel{\wedge}{\text{POM}} = 5\sin 30° = 2.50$

Hence, in algebraic form, the complex number $z = -4.33 - \text{j}2.50$

Exercise 6.2

1) Show, indicating each one clearly, the following complex numbers on a single Argand diagram: $4 + \text{j}3$, $-2 + \text{j}$, $3 - \text{j}4$, $-3.5 - \text{j}2$, $\text{j}3$ and $-\text{j}4$.

2) Find the moduli and arguments of the complex numbers $3 + \text{j}4$ and $4 - \text{j}3$.

3) If the complex number $z_1 = -3 + \text{j}2$ find $|z_1|$ and $\arg z_1$.

4) If the complex number $z_2 = -4 - \text{j}2$ find $|z_2|$ and $\arg z_2$.

5) Express each of the following complex numbers in polar form:
   a) $4 + \text{j}3$ $\qquad\qquad$ **b)** $3 - \text{j}4$ $\qquad\qquad$ **c)** $-3 + \text{j}3$
   d) $-2 - \text{j}$ $\qquad\qquad$ **e)** $\text{j}4$ $\qquad\qquad\qquad$ **f)** $-\text{j}3.5$

6) Convert the following complex numbers, which are given in polar form, into Cartesian form:
   a) $3\ \underline{/45°}$ $\qquad\qquad$ **b)** $5\ \underline{/154°}$ $\qquad\qquad$ **c)** $4.6\ \underline{/-20°}$
   d) $3.2\ \underline{/-120°}$

7) The potential difference across a circuit is given by the complex number $V = 40 + \text{j}35$ volts and the current is given by the complex number $I = 6 + \text{j}3$ amperes. Sketch the appropriate phasors on an Argand diagram and find:
   a) the phase difference (i.e. the angle ϕ) between the phasors for V and I,
   b) the power, given that $\qquad$ power $= |V| \times |I| \times \cos \phi$.

7 Logic Design and Number Systems

Constant – variables – operations NOT, AND and OR – truth tables – Veitch–Kamaugh maps – denary, octal, hexadecimal and binary systems of numbers – octal and binary arithmetic – half-adder and full-adder computer circuits.

INTRODUCTION

This chapter will introduce you to the ideas behind the design of logic circuits which is the way most computer and microprocessors work. You will also become acquainted with number systems – denary, octal, hexadecimal and binary.

We make use of Boolean algebra which, together with the use of binary arithmetic, makes possible our understanding of logic circuitry. To help us handle Boolean expressions we shall use two types of tabular display – namely Truth tables and Vietch–Karnaugh maps.

BOOLEAN ALGEBRA

This topic comprises constants, variables and operations.

A CONSTANT

In normal use a constant is a fixed value or quantity, e.g. 7, 9, ...
In Boolean algebra there are only two constants, namely 0 and 1.
A typical transistor circuit may use 0 volt to represent 0,
and +5 volt to represent 1.

A VARIABLE

In ordinary algebra a variable is a quantity which can change by taking the value of any constant. We use variables x, y ... to represent constants such as 4, 7, $6\frac{1}{2}$ ...

In Boolean algebra variables are denoted by capital letters (italic) such as A, B, X ... but since there are only two possible constants then a variable can only have only one of two values – either 0 or 1.

OPERATIONS

In ordinary algebra we add, subtract, multiply, divide and even change sign when operating on quantities.

In Boolean algebra totally different operations are used. We shall concentrate on three basic operations, namely NOT, AND and OR. There are others but these will suffice at this stage of our studies.

The operations are carried out in electric circuits by electronic units called 'gates', or alternatively by switches in specially designed circuits.

The NOT operation

This is indicated by placing a bar ($\bar{\ }$) over the constant or variable.

Alternatively due to typesetting limitations a solidus (/) may be used before the constant or variable.

Definition of **NOT** operation:	$\bar{1} = 0$	$\bar{0} = 1$

Gate symbol

A —[NOT]— $\bar{A}$

Truth table

Input	Output
A	$\bar{A}$
0	1
1	0

Explanation

When the input signal is 0 then the output is 1, and vice versa.

The AND operation

This is shown by a dot (.) between two constants or variables.

Definition of **AND** operation:	$0.0 = 0$ $0.1 = 0$ $1.0 = 0$ $1.1 = 1$

Gate symbol

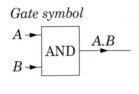

Truth table

Input		Output
A	B	$A.B$
0	0	0
0	1	0
1	0	0
1	1	1

Explanation

The output signal is 1 only when *all* the input signals are 1.

An alternative to the AND logic gate is a circuit containing two switches in series, a closed switch representing 1 and an open switch representing 0.

The OR operation

This is shown by a + symbol between two constants or variables.

Definition of **OR** operation:	$0 + 0 = 0$ $0 + 1 = 1$ $1 + 0 = 1$ $1 + 1 = 1$

Gate symbol

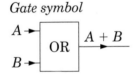

Truth table

Input		Output
A	B	$A + B$
0	0	0
0	1	1
1	0	1
1	1	1

Explanation

The output signal is 1 when one or more of the input signals is 1.

An alternative to the OR logic gate is a circuit containing two switches in parallel, again a closed switch representing 1 and an open switch 0:

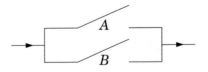

Operating on Constants

We may use the AND with more than one constant, such as $0.1.0.\bar{1}$ and the value of this may be found by taking the terms in pairs – brackets may be used as in ordinary algebra. Thus:

$$0.1.0.\bar{1} = 0.1.0.0 \qquad \text{since} \quad \bar{1} = 0$$
$$= (0.1).(0.0)$$
$$= 0.0 \qquad \text{since} \quad 0.1 = 0 \quad \text{and} \quad 0.0 = 0$$
$$= 0 \qquad \text{since} \quad 0.0 = 0$$

You will see that when constants are ANDed it only needs one 0 or $\bar{1}$ term to make the result zero. Remember that for a 1 result when ANDing all the constants must be 1 or $\bar{0}$.

The diagram opposite shows $0.1.0.\bar{1} = 0$ using two logic gates. Note the four inputs to the AND gate.

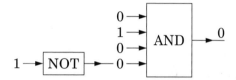

We may also use the OR with more than one constant, such as $0 + 1 + \bar{0}$ and to find the value of this we proceed as before. Thus:

$$0 + 1 + \bar{0} = (0 + 1) + 1 \qquad \text{since} \quad \bar{0} = 1$$
$$= 1 + 1 \qquad \text{since} \quad 0 + 1 = 1$$
$$= 1 \qquad \text{since} \quad 1 + 1 = 1$$

The diagram opposite shows $0 + 1 + 0 = 1$ using two logic gates. Note the three inputs to the OR gate.

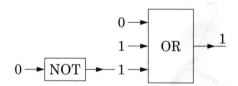

AND and OR may be used but when simplifying an expression AND has preference over OR.

EXAMPLE 7.1

Evaluate $(\bar{1} + 1).1 + 1.\bar{0}$

Now $(\bar{1} + 1).1 + 1.\bar{0}$ = $(0 + 1).1 + 1.1$ since $\bar{1} = 0$ and $\bar{0} = 1$
$= 1.1 + 1.1$ since $0 + 1 = 1$
$= 1 + 1$ since $1.1 = 1$
$= 1$ since $1 + 1 = 1$

Exercise 7.1

Evaluate the following:

1) $0.1.0$

2) $0.1.1.1$

3) $0.\bar{1}.\bar{1}.0$

4) $1 + \bar{1}$

5) $\bar{0} + \bar{0}$

6) $\bar{1} + 0 + 1$

7) $1 + 0.\bar{1}$

8) $\bar{0}.\bar{1} + \bar{0}$

9) $1.0.\bar{1} + \bar{0}.1$

10) $(1 + 0).1$

11) $0.(1 + \bar{1})$

12) $(1 + 1).(\bar{0}.1)$

13) Illustrate questions 1–8 on diagrams with suitable logic gates. Assume that only inputs of 0 and 1 are available.

TRUTH TABLES

You have already met three truth tables which are simply a convenient way of displaying all possible values of input and corresponding output values. They may also be used to find the values of more complicated expressions.

Consider the expression $A.\bar{B} + B.\bar{A}$ whose truth table will cover all the possible combinations of constants 0 and 1 which can be given to variables A and B. Intermediate columns for $\bar{A}$, $\bar{B}$, $A.\bar{B}$ and $B.\bar{A}$ are introduced to make the working out clearer.

A	B	$\bar{A}$	$\bar{B}$	$A.\bar{B}$	$B.\bar{A}$	$A.\bar{B} + B.\bar{A}$
0	0	1	1	0	0	0
0	1	1	0	0	1	1
1	0	0	1	1	0	1
1	1	0	0	0	0	0

As you become more familiar with giving values to variables, like A and B here, you may well find it possible to do the working out mentally and dispense with a truth table – but beware: mistakes are easily made!

EXAMPLE 7.2

Make a truth table for the expression $\bar{A}.B.\bar{C}$

A	B	C	$\bar{A}$	B	$\bar{C}$	$\bar{A}.B.\bar{C}$
0	0	0	1	0	1	0
0	0	1	1	0	0	0
0	1	0	1	1	1	1
0	1	1	1	1	0	0
1	0	0	0	0	1	0
1	0	1	0	0	0	0
1	1	0	0	1	1	0
1	1	1	0	1	0	0

Note the repeat of column B for convenience between $\bar{A}$ and $\bar{C}$.

EXAMPLE 7.3

Use a truth table to verify the identity $A + \bar{A}.B = A + B$

A	B	$\bar{A}$	$\bar{A}.B$	$A + \bar{A}.B$	$A + B$
0	0	1	0	0	0
0	1	1	1	1	1
1	0	0	0	1	1
1	1	0	0	1	1

The last two columns show that, for all possible values of variables A and B, the left hand side of the identity has the same value as the right hand side. Hence the identity $A + \bar{A}.B = A + B$ is verified.

Exercise 7.2

Verify the following identities using suitable truth tables:

1) $A + A.B = A$

2) $A.(A + B) = A$

3) $\overline{(A + B)} = \bar{A}.\bar{B}$

4) $\overline{A.B} = \bar{A} + \bar{B}$

5) $A.(B + C) = A.B + A.C$

6) $A + B.C = (A + B).(A + C)$

SIMPLIFYING BOOLEAN EXPRESSIONS

In the early 1950s Veitch and Karnaugh developed a method of tabulation in the form of a 'map'. We shall call these V–K maps.

Each V–K map contains 2^n squares for n variables.

Thus for 2 variables we have $2^2 = 4$ squares

and   for 3 variables we have $2^3 = 8$ squares

and   for 4 variables we have $2^4 = 16$ squares

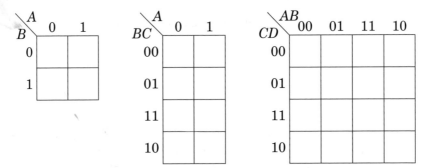

Note the change in order of values 11 and 10 in labelling the squares as compared with the value sequence on a truth table. This is because only one variable changes between adjacent squares.

The following examples will show how to plot on a V–K map.

EXAMPLE 7.4

Plot the expression $A.\overline{B} + B.C$ on a V–K map.

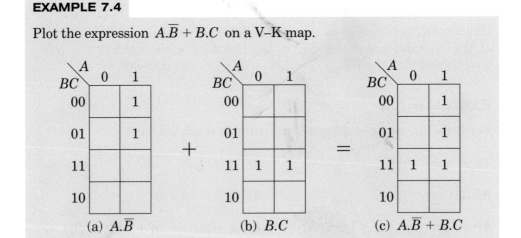

Fig. 7.1

Since there are three variables A, B and C we require an eight square map. First consider $A.\overline{B}$. We shall plot this on its own grid (Fig. 7.1(a)) by putting a 1 in each square where the value of $A.\overline{B}$ equals 1. We can see by inspection that the required 1.1 occurs when $A = 1$ and $B = 0$. The value of the third variable C can be 0 or 1 and does not affect the value of $A.\overline{B}$ so we must plot 1 in squares for both $A = 1$, $B = 0$, $C = 0$, and $A = 1$, $B = 0$, $C = 1$. We say that C is undefined.

Figure 7.1(b) shows the map for $B.C$ whose value is 1 when $B = 1$, $C = 1$ and either $A = 0$ or $A = 1$. Here A is undefined.

Figure 7.1(c) shows the final result of combining (a) and (b).

EXAMPLE 7.5

Plot the expression $A.\overline{B}.C + \overline{C}$ on a V–K map.

We do not need to plot $A.\overline{B}.C$ and $\overline{C}$ on individual maps – so let us look at the whole expression $A.\overline{B}.C + \overline{C}$.   The OR (+) operation means that for a final 1 value then either $A.\overline{B}.C = 1$ or $\overline{C} = 1$, or both.

For $A.\overline{B}.C = 1$ then $A = 1$, $B = 0$ and $C = 1$.

For $\overline{C} = 1$ then $C = 0$ and both A and B are undefined.

The plot is shown in Fig. 7.2.

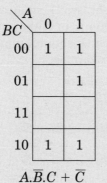

$$A.\overline{B}.C + \overline{C}$$ Fig. 7.2

EXAMPLE 7.6

On a V–K map plot the expression $A.\overline{B}.X + B.Y + \overline{A}.B.X.Y$

This time we have four variables so we need a grid with $2^4 = 16$ squares. Bearing in mind the OR operation, then for the whole expression to be 1 either $A.\overline{B}.X = 1$ for which $A = 1$, $B = 0$ and $X = 1$ and Y is undefined,

or $B.Y = 1$ for which $B = 1$, $Y = 1$ and both A and X are undefined,

or $\overline{A}.B.X.Y = 1$ for which $A = 0$, $B = 1$, $X = 1$ and $Y = 1$.

This last result finds a square already occupied by a 1; this does not matter as long as the square is filled already.

The final result is shown plotted in Fig. 7.3.

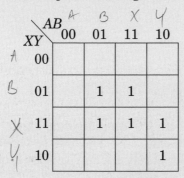

$$A.\overline{B}.X + B.Y + \overline{A}.B.X.Y \qquad \text{Fig. 7.3}$$

Simplified Terms for a V–K Map

These are found from a grouping of two '1 squares' if they are adjacent to each other in the same row or column, or if they are at opposite ends of the same row or column. Also each '1 square' may be used as many times as desired, but must be used at least once.

EXAMPLE 7.7

Use a V–K map to simplify the expression $A.B.C + A.\overline{B}.C$.

Figure 7.4 shows the V–K map of $A.B.C + A.\overline{B}.C$.
Now the two '1 squares' are adjacent and so may be grouped together.
The grouped squares represent:

$$\left.\begin{array}{l} A = 1, \ B = 1, \ C = 1 \\ A = 1, \ B = 0, \ C = 1 \end{array}\right\} \text{ or } A.C \text{ with } B \text{ undefined.}$$

Thus A.C. is a simplified form and we may say that

$$A.B.C + A.\overline{B}.C = A.C$$

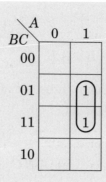

Fig. 7.4

EXAMPLE 7.8

Simplify $X.\overline{Y}.\overline{Z} + \overline{X}.\overline{Y}.Z + X.Y.\overline{Z} + \overline{X}.\overline{Y}.\overline{Z}$ using a V–K map.

Figure 7.5 shows the V–K map of the given expression. There are two pairs of adjacent '1 squares', and one pair comprising '1 squares' at opposite ends of a column. The grouped squares represent:

$$\left.\begin{array}{l} X = 0,\ Y = 0,\ Z = 0 \\ X = 0,\ Y = 0,\ Z = 1 \end{array}\right\} \text{or } \overline{X}.\overline{Y} \text{ with } Z \text{ undefined}$$

$$\left.\begin{array}{l} Y = 0,\ Z = 0,\ X = 0 \\ Y = 0,\ Z = 0,\ X = 1 \end{array}\right\} \text{or } \overline{Y}.\overline{Z} \text{ with } X \text{ undefined}$$

$$\left.\begin{array}{l} X = 1,\ Z = 0,\ Y = 0 \\ X = 1,\ Z = 0,\ Y = 1 \end{array}\right\} \text{or } X.\overline{Z} \text{ with } Y \text{ undefined}$$

Thus $\overline{X}.\overline{Y}. + \overline{Y}.\overline{Z} + X.\overline{Z}$ is a simplified form and we may say that

$$X.\overline{Y}.\overline{Z} + \overline{X}.\overline{Y}.Z + X.Y.\overline{Z} + \overline{X}.\overline{Y}.\overline{Z} = \overline{X}.\overline{Y} + \overline{Y}.\overline{Z} + X.\overline{Z}$$

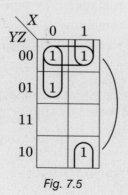

Fig. 7.5

Exercise 7.3

1) Plot the following expressions on a two variable V–K map:

   a) $A + A.B$ **b)** $A + B + A.B$ **c)** $X.\overline{Y} + Y.\overline{X}$

2) Plot the following expressions on a three variable V–K map:

   a) $A + \overline{B} + C$ **b)** $A.\overline{B}.C + \overline{C}$ **c)** $(\overline{A} + B).\overline{C}$

3) Plot the following expressions on a four variable V–K map:

   a) $W.X.Y.Z + W.Y.Z + W.Z$ **b)** $A.B.\overline{C} + \overline{B}.C.D + \overline{A}$

   c) $Z + \overline{W}.\overline{Y}.\overline{X}$

4) Simplify the following expressions using suitable V–K maps:

   a) $A + \overline{A}.B$ **b)** $\overline{A}.\overline{B}.C + \overline{A}.B.C$

   c) $\overline{A}.B.C + A.B.\overline{C} + A.B.C$ **d)** $\overline{A}.B.\overline{C}.D + A.B.\overline{C}.D$

NUMBER SYSTEMS

The Denary System

Let us first consider the ordinary decimal (or denary) system which uses ten digits 0 to 9, and a number base of ten. For example:

$$
\begin{aligned}
2975 &= 2000 + 900 + 70 + 5 \\
&= 2 \times 1000 + 9 \times 100 + 7 \times 10 + 5 \times 1 \\
&= 2 \times 10^3 + 9 \times 10^2 + 7 \times 10^1 + 5 \times 10^0
\end{aligned}
$$

Another way of showing this is a series of divisions:

$$
\begin{aligned}
2975 \div 10 &= 297 \text{ remainder } 5 - \text{units} \\
297 \div 10 &= 29 \text{ remainder } 7 - \text{tens} \\
29 \div 10 &= 2 \text{ remainder } 9 - \text{hundreds} \\
2 \div 10 &= 0 \text{ remainder } 2 - \text{thousands}
\end{aligned}
$$

$$2 \quad 9 \quad 7 \quad 5$$

Why did we choose a ten based system? Probably because we had a total of ten fingers and thumbs and these were our original calculating machine!

The Octal System

Suppose now that the inventor of a number system had lost both his thumbs in a hunting accident – so his counting would have to be done on eight fingers. The octal system has a number base of eight, and uses eight digits 0 to 7.

So let us use the series of divisions method to convert 2975 to octal:

$$2975 \div 8 = 371 \text{ remainder } 7 \quad - \quad \text{units}$$
$$371 \div 8 = 46 \text{ remainder } 3 \quad - \quad \text{eights}$$
$$46 \div 8 = 5 \text{ remainder } 6 \quad - \quad (8 \times 8)\text{s}$$
$$5 \div 8 = 0 \text{ remainder } 5 \quad - \quad (8 \times 8 \times 8)\text{s}$$

$$5 \quad 6 \quad 3 \quad 7$$

So the octal number 5637 represents $\Big\}\ 5 \times 8^3 + 6 \times 8^2 + 3 \times 8^1 + 7 \times 8^0$

Thus $\qquad\qquad$ 2975 (denary) $=$ 5637 (octal)

Or using notation $\qquad\qquad 2975_{10} = 5637_8$

EXAMPLE 7.9

Convert 71 293 to octal.

$$71\,293 \div 8 = 8911 \text{ remainder } 5 \quad \text{units}$$
$$8911 \div 8 = 1113 \text{ remainder } 7 \quad \text{eights}$$
$$1113 \div 8 = 139 \text{ remainder } 1 \quad (8^2)\text{s}$$
$$139 \div 8 = 17 \text{ remainder } 3 \quad (8^3)\text{s}$$
$$17 \div 8 = 2 \text{ remainder } 1 \quad (8^4)\text{s}$$
$$2 \div 8 = 0 \text{ remainder } 2 \quad (8^5)\text{s}$$

Thus $\quad 71\,293_{10} = 2 \times 8^5 + 1 \times 8^4 + 3 \times 8^3 + 1 \times 8^2 + 7 \times 8^1 + 5$

or $\qquad 71\,293_{10} = 213\,175_8$

EXAMPLE 7.10

Convert $174\,023_8$ to denary.

$$174\,023_8 = 1 \times 8^5 + 7 \times 8^4 + 4 \times 8^3 + 0 \times 8^2 + 2 \times 8^1 + 3$$
$$= 32\,768 + 28\,672 + 2048 + 0 + 16 + 3 = 63\,507$$

So $174\,023_8 = 63\,507_{10}$

Octal arithmetic

Counting in octal: $0,\ 1,\ 2,\ 3,\ 4,\ 5,\ 6,\ 7,\ ?$

There is no 8 so we continue with 10 meaning 1 eight and no units:

$$0, 1, 2, 3, 4, 5, 6, 7, 10, 11, 12, 13, 14, 15, 16, 17,$$
$$20, 21, 22, 23, \dots 75, 76, 77, ?$$

There is no 78 or 80 and we move to 100 which means one $8^2 = 64$ and no eights and no units, giving $76,\ 77,\ 100,\ 101,\ 102,\ \dots$

Adding: $1\ 5\ 6\ 7_8$ Working from R.H. column to the left, then

$+$ $1\ 5\ 4_8$

$\underline{\quad 1\ 1 \leftarrow \text{carries}}$ $7 + 4\ =\ 13_8$ giving 3 carry 1

$1\ 7\ 4\ 3_8$ $6 + 5 + 1\ =\ 14_8$ giving 4 carry 1

$5 + 1 + 1\ =\ 7_8$ giving 7 etc.

The Hexadecimal System

This is a number system with a base of sixteen. It may be used, as maybe also the octal system, in computer arithmetic in conjunction with the binary system which we shall meet next.

The hexadecimal system has a base of sixteen and thus needs sixteen digits. We have only ten digits in common use, namely 0 to 9, so we will have to invent another six – let us call these A, B, C, D, E and F. So counting in hexadecimal will be:

$$0, 1, 2, 3, 4, 5, 6, 7, 8, 9, A, B, C, D, E, F, 10, 11, 12, 13,$$
$$14, 15, 16, 17, 18, 19, 1A, 1B, 1C, 1D, 1E, 1F, 20, 21$$
$$23, 24, \dots$$

We can change a hexagonal number to an ordinary denary number, and vice versa, by similar methods to those used for the octal system but using, of course, 16 instead of 8.

The Binary System

This is a number system using a base of two – this means that only two digits are needed, namely 0 and 1. You will immediately note, no doubt, why binary numbers are favoured in computer arithmetic as these are the two constants used in Boolean algebra logic.

Let us use the series of divisions method to convert 45 to binary:

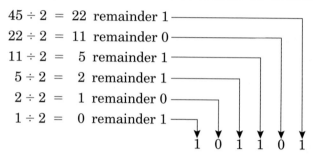

$$45 \div 2 = 22 \text{ remainder } 1$$
$$22 \div 2 = 11 \text{ remainder } 0$$
$$11 \div 2 = 5 \text{ remainder } 1$$
$$5 \div 2 = 2 \text{ remainder } 1$$
$$2 \div 2 = 1 \text{ remainder } 0$$
$$1 \div 2 = 0 \text{ remainder } 1$$

$$1 \quad 0 \quad 1 \quad 1 \quad 0 \quad 1$$

So the binary number 101101 represents:

$$1 \times 2^5 + 0 \times 2^4 + 1 \times 2^3 + 1 \times 2^2 + 0 \times 2^1 + 1 \times 2^0$$

or $\qquad 1 \times 32 + 0 \times 16 + 1 \times 8 + 1 \times 4 + 0 \times 2 + 1 \times 1$

or $\qquad 32 + 8 + 4 + 1 = 45$

Thus $\qquad\qquad\qquad$ 45 denary $= 101101$ binary

or $\qquad\qquad\qquad\qquad 45_{10} = 101101_2$

Bicimals

In the denary system figures to the right of the decimal point are called decimals. In the binary system figures to the right of the bicimal point are called bicimals.

So the binary number 101.01101 represents

$$1 \times 2^2 + 0 \times 2^1 + 1 \times 2^0 + 0 \times 2^{-1} + 1 \times 2^{-2} + 1 \times 2^{-3} + 0 \times 2^{-4} + 1 \times 2^{-5}$$

or $\qquad 1 \times 4 + 0 \times 2 + 1 \times 1 + 0 \times \frac{1}{2} + 1 \times \frac{1}{4} + 1 \times \frac{1}{8} + 0 \times \frac{1}{16} + 1 \times \frac{1}{32}$

or $\qquad 4 + 1 + \frac{1}{4} + \frac{1}{8} + \frac{1}{32}$

or $\qquad 4 + 1 + 0.25 + 0.125 + 0.031\,25$

or $\qquad 5.40625$

Thus   101.01101 binary = 5.406 25 denary

or $101.01101_2 = 5.406\,25_{10}$

To convert decimals 0.406 25 to bicimal we use a series of multiplications:

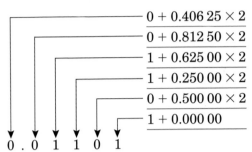

$0 + 0.406\,25 \times 2$ Note that in each case it
$0 + 0.812\,50 \times 2$ is only the decimal
$1 + 0.625\,00 \times 2$ figures which are
$1 + 0.250\,00 \times 2$ multiplied by 2.
$0 + 0.500\,00 \times 2$
$1 + 0.000\,00$ As soon as a whole
                              number figure 1 appears
0 . 0 1 1 0 1   it is shown separately
                              and *not* multiplied by 2.

Thus $0.406\,25_{10} = 0.01101_2$

Binary arithmetic

An understanding of binary arithmetic is useful since most computer circuits perform calculations using binary.

We shall show how to add and multiply in binary and then finish by looking at basic arithmetic logic circuits.

EXAMPLE 7.11

Add 10101_2 and 1111_2. Working from the RH column to the left, then:

```
    1 0 1 0 1
  +   1 1 1 1
    1 1 1 1← carries
    ─────────
    1 0 0 1 0 0₂
```

$1 + 1 = 10$ giving 0 carry 1
$0 + 1 + 1 = 10$ giving 0 carry 1
$1 + 1 + 1 = 11$ giving 1 carry 1

and so on.

EXAMPLE 7.12

Multiply 10011_2 by 1101_2.

```
          1 0 0 1 1
    ×       1 1 0 1
        ─────────────
          1 0 0 1 1    (a)
        0 0 0 0 0      (b)
      1 0 0 1 1        (c)
    1 0 0 1 1          (d)
    ─────────────
```
Sum of lines a, b, c and d.

The computer would not add the four lines a, b, c and d in one go, but would probably proceed in stages:

```
      1 0 0 1 1        (a)
    + 0 0 0 0 0        (b)
      1 0 0 1 1        (a) + (b)
    + 1 0 0 1 1        (c)
    1 0 1 1 1 1 1      (a) + (b) + (c)
  + 1 0 0 1 1          (d)
        1 1            ← carries
  1 1 1 1 0 1 1 1₂     Answer: sum of a, b, c and d.
```

BASIC ARITHMETIC COMPUTER CIRCUITS

The basic arithmetic circuit is the ADDER. This will enable addition and multiplication (which, as you have seen, is merely successive addition) to be performed. There are two variants, namely the HALF-ADDER and the FULL-ADDER.

The Half-adder (Two Input)

This circuit (Fig. 7.6) has two inputs and two outputs. The inputs are two binary digits and the outputs are the sum and carry from binary addition – the corresponding truth table is also shown.

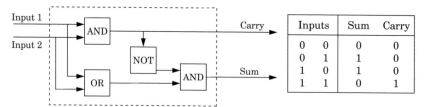

Inputs		Sum	Carry
0	0	0	0
0	1	1	0
1	0	1	0
1	1	0	1

Fig. 7.6   Half-adder circuit

The Full-adder (Three Input)

Here we have a circuit (Fig. 7.7) with three inputs – two binary and a 'carry in' from a previous stage. Again the outputs are the sum and 'carry out' from binary addition.

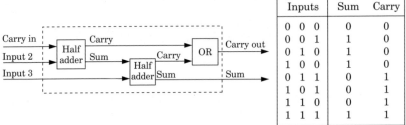

Inputs			Sum	Carry
0	0	0	0	0
0	0	1	1	0
0	1	0	1	0
1	0	0	1	0
0	1	1	0	1
1	0	1	0	1
1	1	0	0	1
1	1	1	1	1

Fig. 7.7   Full-adder circuit

Exercise 7.4

1) Convert to denary
   a) 63_8 b) 217_8 c) 463_8

2) Convert to octal
   a) 29_{10} b) 243_{10} c) 1796_{10}

3) Find the sum of
   a) 637_8 and 56_8 b) 4360_8 and 1234_8

4) Check the results of question 3 by converting each number to denary, adding, and then converting back to octal.

5) Convert to binary
   a) 23_{10} b) 125_{10} c) 97_{10}

6) Convert to denary
   a) 10110   b) 111001   c) 1011010

7) Convert to denary
   a) 0.1101   b) 0.0111   c) 0.0011

8) Convert to binary
   a) $\frac{3}{8}$ b) $\frac{5}{16}$ c) $\frac{7}{8}$

9) Convert to binary correct to seven bicimal places:
   a) 0.169   b) 18.467   c) 108.710
   Hint: In **b)** and **c)** convert the whole numbers part separately from the decimal parts and then combine the results for the final answer.

10) Add the following binary numbers:
   a) $1011 + 11$ b) $11011 + 1011$
   c) $10111 + 11010 + 111$

11) Multiply the following binary numbers:
   a) 101×111 b) 1011×1010 c) 11011×1101

Cartesian and Polar Co-ordinates

8

Define polar co-ordinates – convert from Cartesian to polar co-ordinates and vice versa – plot graphs of functions defined in polar form

---

INTRODUCTION

If you have worked through the chapter on complex numbers you will have met already the idea of polar form from using the Argand diagram. Here we extend the idea to plotting graphs in polar form – as you will see it can be advantageous in some instances.

POLAR COORDINATES

You have met previously Cartesian (or rectangular) coordinates with which P may be given as the point (x, y) as shown in Fig. 8.1.

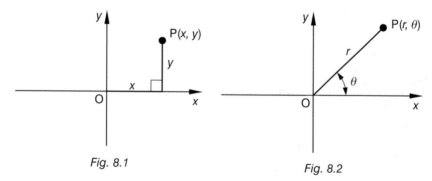

Fig. 8.1

Fig. 8.2

Another way of giving the location of P is by using its distance r from the origin O, together with angle θ that OP makes with the horizontal axis OX (Fig. 8.2). Figure 8.3 shows five typical points, plotted on a polar grid, together with their respective polar coordinates.

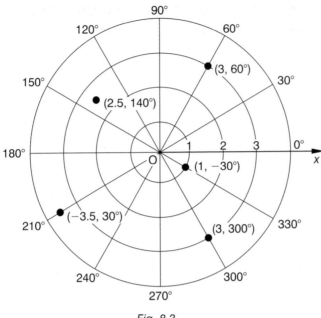

Fig. 8.3

Positive values of θ are always measured anticlockwise from OX whilst negative values are measured clockwise (see Fig. 8.3): you will not often meet the latter case. A negative value of r means that the 'radius length' is extended 'backwards' through O from the normal angle position: see point $(-3.5, 30°)$ in Fig. 8.3. Note that in polar coordinates it is possible to define a point in more than one way: for example the point $(3, 300°)$ in Fig. 8.3 could also be defined as $(-3, 120°)$.

Polar graph paper is available but is not so readily obtained as the common linear variety. However, it should be sufficient here for you to sketch the polar curves we meet, or perhaps draw your own simple polar grid similar to that in Fig. 8.3.

A typical practical application of a polar plot would be to values of luminous intensity of an electric light bulb, to show how it varied around the bulb relative to the position of the filament etc.

RELATIONSHIP BETWEEN CARTESIAN AND POLAR COORDINATES

From the right-angled triangle POM in Fig. 8.4 we can see that:

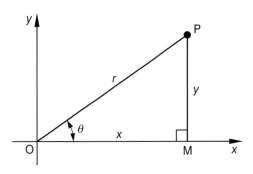

$$x = r \cos \theta$$

and $\quad y = r \sin \theta$

Also $\quad r = \sqrt{x^2 + y^2}$

and $\quad \tan \theta = \dfrac{y}{x}$

Fig. 8.4

Using the above relationships it is reasonably easy to convert from Cartesian to polar coordinates, and vice versa. Always make a sketch of the problem because this will enable you to see which quadrant you are dealing with.

EXAMPLE 8.1

Find the polar coordinates of the point $(-4, 3)$.

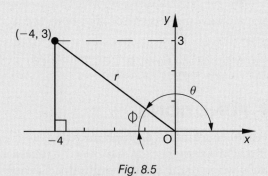

Fig. 8.5

From Fig. 8.5 $\qquad \tan \phi = \dfrac{3}{4} = 0.75$

$\therefore \qquad\qquad\qquad \phi = 36.9°$

Thus $\qquad\qquad \theta = 180° - 36.9° = 143.1°$

Also $\qquad\qquad r = \sqrt{3^2 + 4^2} = 5$

Thus the point is $(5, 143.1°)$ in polar form.

EXAMPLE 8.2

Express in Cartesian form the point $(6, 231°)$.

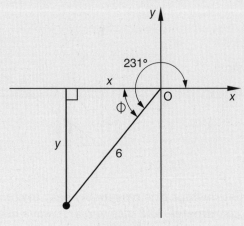

Fig. 8.6

From Fig. 8.6 $\phi = 231° - 180° = 51°$

Thus $x = 6\cos 51° = 3.78$

and $y = 6\sin 51° = 4.66$

Now, having drawn a diagram, we can see that both the x and y coordinates are negative.

Thus the Cartesian form of the point is $(-3.78, -4.66)$

GRAPHS OF FUNCTIONS

When using Cartesian coordinates a graph may be drawn to illustrate y as a function of x. For example $y = mx + c$ represents a straight line graph. Similarly when using polar coordinates a graph may be drawn to illustrate r in terms of θ. In the examples which follow we will sketch some of the more common polar graphs and this should enable us to become familiar with their shapes.

EXAMPLE 8.3

Sketch the graph of $r = \sin \theta$ between $\theta = 0°$ and $\theta = 360°$.

From experience we know that $\sin \theta$ (and hence r) has a maximum value of $+1$, and a minimum value of -1. This will help initially in labelling our polar grid (Fig. 8.7). We may plot the graph from values obtained from our scientific calculator. We first measure off the angle θ and then measure off the length r along the angle boundary line.

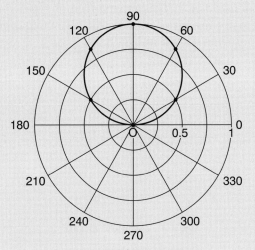

Fig. 8.7   Graph of $r = \sin \theta$

Plotting from $\theta = 1$ to $\theta = 180°$ is straightforward. However, between $180°$ and $360°$ the values of r are negative and we merely repeat the curve already plotted. This curve is, in fact, a circle.

For interest you may like to verify that this is the only curve for *any* value of θ however large – both positive and negative.

EXAMPLE 8.4

Sketch the graph of $r = 2$.

Here no mention is made of the angle θ. This means that $r = 2$ defines the graph whatever value is given for θ. Thus the graph is a circle of radius 2 as shown in Fig. 8.8.

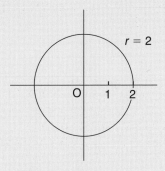

Fig. 8.8   Graph of $r = 2$

EXAMPLE 8.5

Sketch the graph of $\theta = 40°$.

Here no mention is made of r. This means that $\theta = 40°$ defines the graph whatever value is given for r (whether positive or negative). Hence the graph is a straight line as shown in Fig. 8.9.

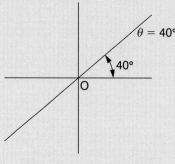

Fig. 8.9   Graph of $\theta = 40°$

EXAMPLE 8.6

Sketch the graph of $r = 3\theta$ for angle values equivalent to the range $0°$ to $540°$.

As is usual in mathematics when an angle is used directly in calculations its value must be in radians. But we find it more convenient to plot the angle values using degrees and so we must be careful to find the corresponding angle values in radians when calculating r values.

Your scientific calculator may convert directly from degrees to radians.

The graph is known as an Archimedean spiral and is shown in Fig. 8.10.

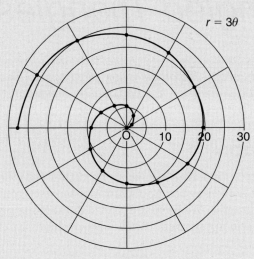

$r = 3\theta$

Fig. 8.10   Graph of r = 3θ

Exercise 8.1

1) Find the polar coordinates of the following points:

    a) $(5, 7)$ **b)** $(-2, 3)$ **c)** $(2, -3)$ **d)** $(-3, -4)$

2) Find the Cartesian coordinates of the following points:

    a) $(2, 35°)$ **b)** $(3, 127°)$ **c)** $(1.5, 240°)$

    d) $(0.6, 312°)$ **e)** $(2.3, -21°)$ **f)** $(-5, 130°)$

Sketch the graphs for the polar equations in the questions which follow:

3) $r = 2\cos\theta$ **4)** $r = 3$ **5)** $\theta = 120°$

6) $r = \theta$ **7)** $r = \sin^2\theta$ **8)** $r = \cos^2\theta$

9) $r = 3\sin 2\theta$ **10)** $r = \cos 3\theta$ **11)** $r = a(1 + \cos\theta)$

9 Trigonometry Right Angled Triangles

Basic trigonometric ratios – reciprocal ratios – Pythagoras theorem – application to solving problems – fine engineering measurements – screw threads – belt drives – frameworks.

INTRODUCTION

Many problems in engineering appear to be rather complicated but, on inspection, can be solved by the application of basic trigonometrical ideas to right angled triangles. We have met these definitions previously so we will review these and apply them to relevant problems.

BASIC TRIGONOMETRIC RATIOS

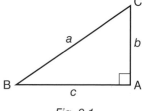

Using the conventially labelled right angle triangle (Fig. 9.1) our three basic definitions are:

Fig. 9.1

$$\sin B = \frac{b}{a} \qquad \cos B = \frac{c}{a} \qquad \tan B = \frac{b}{c}$$

The three reciprocal ratios are defined as follows:

$$\operatorname{cosec} B = \frac{1}{\sin B} \qquad \sec B = \frac{1}{\cos B} \qquad \cot B = \frac{1}{\tan B}$$

Also the theorem of Pythagoras gives:

$$a^2 = b^2 + c^2$$

ACCURATE MEASUREMENTS IN MANUFACTURE

Here we have the application of basic trigonometric functions to problems faced by the fine measurement engineer or perhaps the production quality control inspector.

The Sine Bar

A sine bar is used for measurements which require the angle to be determined closer than $5'$. The method of using the instrument is shown in Fig. 9.2. The distance l is usually made $100\,mm$ or $200\,mm$ to facilitate calculations. h is the difference in height between the two rollers and $h = l\sin\theta$.

EXAMPLE 9.1

The angle of $15°42'$ is to be checked on the metal block shown in Fig. 9.2. Find the difference in height between the two rollers which support the ends of the $200\,mm$ sine bar.

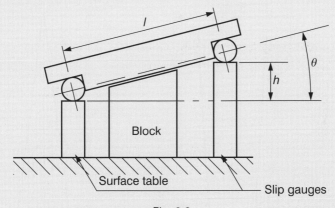

Fig. 9.2

Now $h = l\sin\theta = 200 \times \sin 15°42' = 200 \times 0.2706 = 54.12\,mm$

The difference in height of the slip gauges must therefore be $54.12\,mm$ if the angle is correct.

Reference Rollers and Balls

Sets of rollers can be obtained which are guaranteed to be within 0.002 mm for both diameter and roundness. By using rollers and balls many problems in measurement can be solved, some examples of which are given below.

EXAMPLE 9.2

A taper piece has a taper of 1 in 8 on the diameter. Two pairs of rollers 15.00 mm in diameter are used to check the taper as shown in Fig. 9.3. The measurement over the top rollers is 55.87 mm. Find:

a) the measurement over the bottom rollers if the taper is correct,
b) the bottom diameter of the job.

The first step is to find the angle of the taper.

Using Fig. 9.4 we see that $\tan \alpha = \dfrac{0.5}{8}$ so $\alpha = 3°34'$.

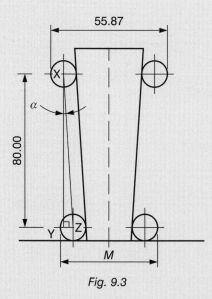

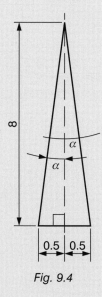

Fig. 9.3   Fig. 9.4

a) Referring to Fig. 9.3, we have in $\triangle XYZ$ that:

$$\angle \alpha = 3°34', \quad XY = 80.00 \text{ mm}$$

but $$\dfrac{YZ}{XY} = \tan \alpha$$

$\therefore$ $$YZ = 80 \times \tan 3°34' = 5.00$$

Thus $$M = 55.87 - 2 \times 5.00 = 45.87 \text{ mm}$$

b) Referring to Fig. 9.5, we need to find AB.

$$\angle ABC = 90° - 3°34' = 86°26'$$

Since AB and BC are tangents, the line BZ bisects $\angle ABC$.

Hence $\qquad\qquad$ ABZ $= 43°13'$

In $\triangle ABZ \qquad \dfrac{AB}{AZ} = \cot 43°13'$

$\therefore \qquad\qquad$ AB $= AZ(\cot 43°13')$

$\qquad\qquad\qquad\quad = 7.5(\cot 43°13') = 7.982 \text{ mm}$

$\therefore \quad$ Bottom diameter $= M - (2 \times AB) - (2 \times \text{radius of roller})$

$\qquad\qquad\qquad\qquad = 45.87 - 2 \times 7.982 - 2 \times 7.50 = 14.91 \text{ mm}$

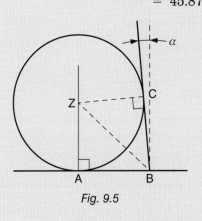

Fig. 9.5

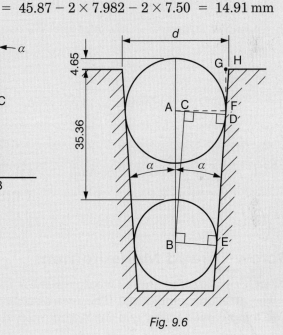

Fig. 9.6

EXAMPLE 9.3

The tapered hole shown was inspected by using two balls 25.00 and 20.00 mm diameter, respectively. The measurements indicated in Fig. 9.6 were obtained. Find:

a) the included angle of taper 2α,
b) the top diameter d of the hole.

a) In Fig. 9.6, A and B are the centres of the balls and E and D are points where the balls just touch the sides of the hole.

$$\angle ADE = \angle BED = 90° \quad \text{(angles between radius and tangent)}$$

Draw BC parallel to DE; then in $\triangle ABC$

$$AC = 12.50 - 10.00 = 2.50 \text{ mm}$$
$$AB = 35.36 + 4.65 - 12.50 + 10.00 = 37.51 \text{ mm}$$

Now $\qquad \angle ACB = 90° \quad \text{and} \quad \angle ABC = \alpha$

$$\sin \alpha = \frac{AC}{AB} = \frac{2.50}{37.51} \qquad \therefore \quad \alpha = 3°49'$$

$\therefore \qquad$ Included angle of taper $= 2\alpha = 2 \times 3°49' = 7°38'$.

b) To find d, draw AF horizontal and FG vertical.

Then $\qquad\qquad\qquad d = 2 \times (AF + GH)$

In $\triangle AFD$,

$$AD = 12.50 \text{ mm}, \quad \angle FAD = \alpha, \quad \angle ADF = 90°$$

$\therefore \qquad AF = AD \sec \alpha = 12.50 \times \sec 3°49' = 12.53 \text{ mm}$

In $\triangle GFH$,

$GF = 12.50 - 4.65 = 7.85 \text{ mm}, \quad \angle FGH = 90°, \quad \angle GFH = \alpha$

$\therefore \qquad GH = GF \tan \alpha = 7.85 \times \tan 3°49' = 0.52 \text{ mm}$

Thus $\qquad d = 2(AF + GH) = 2 \times (12.53 + 0.52) = 26.10 \text{ mm}$

Screw Thread Measurement

When accurate screw thread measurement is required the method of 2 or 3 wire measurement is used. The 3-wire method is used when checking with a hand micrometer and the 2-wire method is used with bench

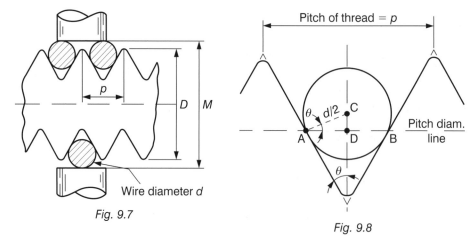

Fig. 9.7

Fig. 9.8

micrometers. The methods are fundamentally the same and allow the pitch or effective diameter to be measured (Fig. 9.7).

For the most accurate results each type and size of thread requires wires of a certain size. The best wire size is one which just touches the thread flanks at the pitch diameter as shown in Fig. 9.8.

At the pitch line diameter the distance between the flanks of the thread is equal to half the pitch.

Hence in Fig. 9.8 $AB = \dfrac{p}{2}$ and $AD = \dfrac{p}{4}$

In $\triangle ADC$, $\dfrac{AC}{AD} = \sec \theta$

$\therefore$ $AC = AD \times \sec \theta$

$\therefore$ $\dfrac{d}{2} = \dfrac{p}{4} \sec \theta \quad \therefore \quad d = \dfrac{p}{2} \sec \theta$

For a metric thread, $\theta = 30°$ and hence the best wire size is

$$d = \dfrac{p}{2} \sec 30° = 0.5774p$$

Formula for Checking the Form of a Metric Thread

Although the best wire size should be used, in practice the wire used will vary a little from this best size. In order to determine the distance over the wires for a specific thread (M in Fig. 9.7) the following formula is used:

$$M = D - \dfrac{5p}{6} \cot \theta + d(\operatorname{cosec} \theta + 1)$$

For a metric thread, the formula becomes

$$M = D - \dfrac{5p}{6} \cot 30° + d(\operatorname{cosec} 30° + 1)$$

$$= D - 1.4434p + d(2 + 1)$$

$$= D - 1.4434p + 3d$$

EXAMPLE 9.4

A metric thread having a major diameter of 20 mm and a pitch of 2.5 mm is to be checked using the best wire size for this particular thread. Find:

a) the best wire size,
b) the measurement over the wires if the thread is correct.

a) The best wire size is
$$d = 0.5774p = 0.5774 \times 2.5 = 1.4435 \text{ mm}$$

b) The measurement over the wires is
$$M = D - 1.4434p + 3d$$
$$= 20 - 1.4434 \times 2.5 + 3 \times 1.4435 = 20.722 \text{ mm}$$

Exercise 9.1

1) Calculate the setting of a 100 mm sine bar to measure an angle of 27°15′.

2) Calculate the setting of a 200 mm sine bar to check a taper of 1 in 8 on diameter.

3) A steel ball 40 mm in diameter is used to check the taper hole, a section of which is shown in Fig. 9.9. If the taper is correct, what is the dimension x?

4) A taper plug gauge is being checked by means of reference rollers and slip gauges. The set-up is as shown in Fig. 9.10. Find the included angle of the taper of the gauge and also the top and bottom diameters.

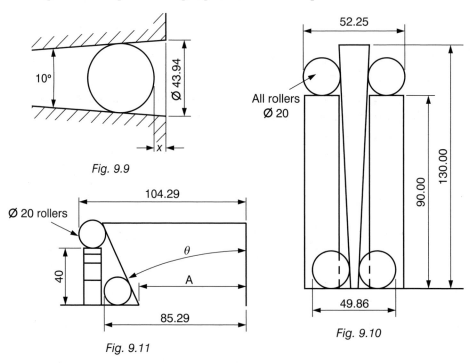

Fig. 9.9

Fig. 9.11

Fig. 9.10

5) Figure 9.11 shows a dovetail being checked by rollers and slip gauges. Find the angle θ and the dimension A.

6) Calculate the dimension M which is needed for checking the groove, a cross-section of which is shown in Fig. 9.12.

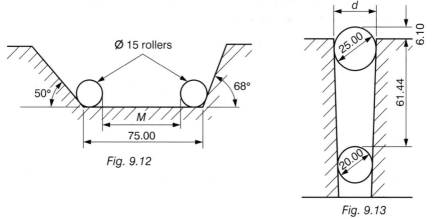

Fig. 9.12

Fig. 9.13

7) A tapered hole has a maximum diameter of 32.00 mm and an included angle of 16°. A ball having a diameter of 20 mm is placed in the hole. Calculate the distance between the top of the hole and the top of the ball.

8) Figure 9.13 shows the dimensions obtained in checking a tapered hole. Find the included angle of taper of the hole and the top diameter d.

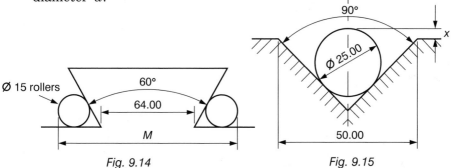

Fig. 9.14

Fig. 9.15

9) Find the checking dimension M for the symmetrical dovetail slide shown in Fig. 9.14.

10) Figure 9.15 shows a Vee block being checked by means of a reference roller. If the block is correct what is the dimension x?

11) A metric thread having a major diameter of 52 mm and a pitch of 5 mm is to be checked by the 3-wire method. Determine the best size of wire and, using the best wire size, determine the measurement over the wires if the thread is correct.

12) A metric thread having a major diameter of 30 mm and a pitch of 2 mm is to be checked using wires whose diameters are 1.14 mm. Calculate the measurement over the wires that will be obtained if the thread is correct.

LENGTHS OF BELTS

Although gears are usually preferred for power transmission, belts (generally Vee belts) are used in applications where a more flexible drive is needed since they absorb and smooth out shock loading.

There are two distinct cases, open belts and crossed belts, as shown in Fig. 9.16.

> With the open belt, pulleys revolve in the *same* direction.
>
> With the crossed belt, pulleys revolve in *opposite* directions.

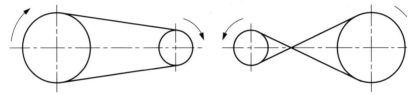

Fig. 9.16

EXAMPLE 9.5

Find the length of an open belt which passes over two pulleys of 200 mm and 300 mm diameter, respectively. The distance between the pulley centres is 900 mm.

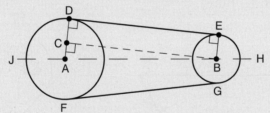

Fig. 9.17

Referring to Fig. 9.17 the total length of the belt is made up of the two straight lengths DE and FG and the arcs DJF and EHG.

$$\angle ADE = \angle BED = 90° \quad \text{(angle between radius and tangent)}$$

Draw CB parallel to DE. Then in $\triangle ABC$,

$$AC = AD - EB = 150 - 100 = 50 \text{ mm}$$
$$AB = 900 \text{ mm (given)}, \quad \angle ACB = 90°$$

$$\therefore \quad \cos CAB = \frac{AC}{AB} = \frac{50}{900} = 0.0556 \quad \therefore \quad \angle CAB = 86°49'$$

Now $\qquad BC = AB \sin CAB = 900 \times \sin 86°49' = 898.6 \text{ mm}$

Also $\qquad \angle EBH = \angle CAB = 86°49'$

Hence the arc EHG subtends an angle of $2 \times 86°49' = 137°38'$ at the centre.

$$\text{Length of arc EHG} = 2\pi \times 100 \times \frac{173°38'}{360°} = 303.0 \text{ mm}$$

Now $\angle\text{DAJ} = 180° - \angle\text{CAB} = 180° - 86°49' = 93°11'$

The arc DJF therefore subtends an angle of $2 \times 93°11' = 186°22'$ at the centre.

$$\text{Length of arc DJF} = \frac{2\pi \times 150 \times 186°22'}{360°} = 488.0\,\text{mm}$$

∴   Total length of belt $= 2 \times \text{BC} + \text{Arc EHG} + \text{Arc DJF}$
$$= 2 \times 898.6 + 303.0 + 488.0 = 2588\,\text{mm}$$

EXAMPLE 9.6

Two pulleys 200 mm and 300 mm in diameter respectively are placed 1200 mm apart. They are connected by a closed belt. Find the length of the belt required.

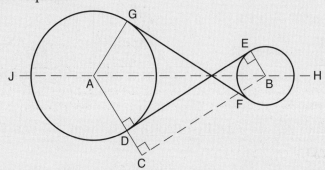

Fig. 9.18

The length of the belt is made up of two straight lengths ED and FG, arc GJD and arc EHF, as shown in Fig. 9.18.

$$\angle\text{ADE} = \angle\text{DEB} = 90° \quad \text{(angle between radius and tangent)}$$

Draw CB parallel to DE; then in $\triangle\text{ABC}$

$$\text{AC} = 100 + 150 = 250\,\text{mm}, \quad \text{AB} = 1200\,\text{mm}, \quad \angle\text{ACB} = 90°,$$

∴ $\cos\text{CAB} = \dfrac{\text{AC}}{\text{AB}} = \dfrac{250}{1200} = 0.2083 \quad \therefore \quad \angle\text{CAB} = 77°59'$

Thus $\text{BC} = \text{AB} \times \sin\text{CAB} = 1200 \times \sin 77°59' = 1174\,\text{mm}$

Also $\angle\text{EBA} = \angle\text{CAB} = 77°59'$

∴ $\angle\text{EBH} = 180° - 77°59' = 102°1'$

The arc EHF therefore subtends an angle $2 \times 102°1'$ at the centre.

∴ $\text{Length of arc EHF} = 2\pi \times 100 \times \dfrac{204°2'}{360°} = 356\,\text{mm}$

Similarly

$$\text{Length of arc GJD} = 2\pi \times 150 \times \frac{204°2'}{360°} = 534\,\text{mm}$$

∴   Total length of belt $= 2 \times 1174 + 356 + 534 = 3238\,\text{mm}$

FRAMEWORKS

Frameworks hardly need an introduction as we can all think of situations where they are used in engineering, although most of these would be for building and construction work. Typical mechanical engineering uses are in crane frameworks and motor vehicle construction.

EXAMPLE 9.7

Figure 9.19 shows a framework. Calculate the lengths of the members BC, BD and AC.

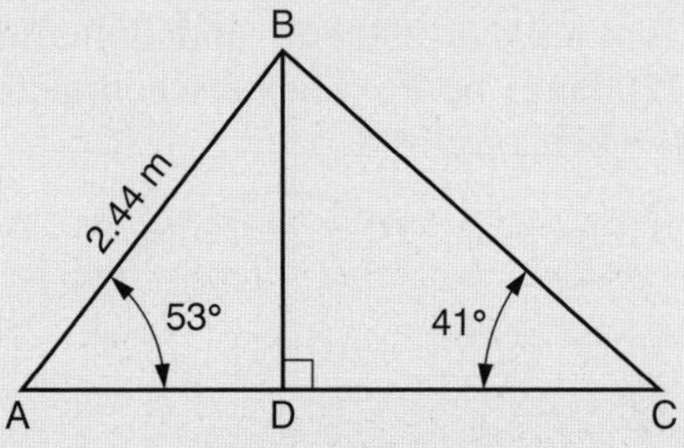

In $\triangle ABD$, $\dfrac{BD}{AB} = \sin 53°$

$\quad BD = AB \sin 53°$

$\qquad = 2.44 \sin 53° = 1.95$

In $\triangle BCD$, $\dfrac{BC}{BD} = \operatorname{cosec} 41°$

$\quad BC = BD \operatorname{cosec} 41°$

$\qquad = 1.95 \operatorname{cosec} 41° = 2.97$

Fig. 9.19

To find the length of AC we must first find the lengths of AD and DC.

In $\triangle ABD$, $\dfrac{AD}{AB} = \cos 53°$

$\therefore \qquad AD = AB \cos 53° = 2.44 \cos 53° = 1.47$

In $\triangle BCD$, $\dfrac{DC}{BC} = \cos 41°$

$\therefore \qquad DC = BC \cos 41° = 2.97 \cos 41° = 2.24$

Thus $\qquad AC = AD + DC = 1.47 + 2.24 = 3.71$

Hence BD is 1.95 m, BC is 2.97 m and AC is 3.71 m.

Exercise 9.2

1) A belt passes over a pulley 1200 mm in diameter. The angle of contact between the pulley and the belt is 230°. Find the length of belt in contact with the pulley.

2) An open belt passes over two pulleys 900 mm and 600 mm in diameter, respectively. If the centres of the pulleys are 1500 mm apart, find the length of the belt required.

3) Two pulleys of diameters 1400 mm and 900 mm, respectively, with centres 4.5 m apart, are connected by an open belt. Find its length.

4) An open belt connects two pulleys of diameters 120 mm and 300 mm with centres 300 mm apart. Calculate the length of the belt.

5) A crossed belt passes over two pulleys each of 450 mm diameter. If their centres are 600 mm apart, calculate the length of the belt.

6) If, in question 2, a crossed belt is used, what will be its length?

7) A crossed belt passes over two pulleys 900 mm and 1500 mm in diameter, respectively, which have their centres 6 m apart. Find its length.

8) A crossed belt passes over two pulleys, one of 280 mm diameter and the other of 380 mm diameter. The angle between the straight parts of the belt is 90°. Find the length of the belt.

9) A framework, to be used horizontally for supporting a machine tool lifting device, is shown in Fig. 9.20 as triangle ABC. Calculate the length AC.

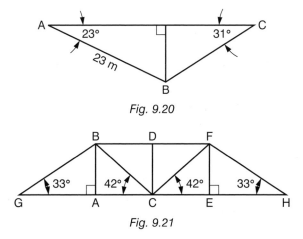

Fig. 9.20

Fig. 9.21

10) The framework shown in Fig. 9.21 has been designed for supporting an overhead gantry rail type transfer system for use in the construction of heavy earth moving equipment. In order to assess the total length of material required, and also the individual lengths, you have been asked to calculate the lengths of all the members. You have been told that AB = CD = EF = 2.4 m.

10 Trigonometry Scalene Triangles

The sine rule – the cosine rule – area of a triangle – circumscribing circle.

INTRODUCTION

Remember the name 'scalene'? It is the mathematical name for 'any old triangle' – a triangle in which all the sides and angles are of different sizes. You will, no doubt, have studied this topic previously so working through the examples will be good revision in solving triangles.

THE SINE, COSINE, AND AREA FORMULAE

We have met previously the *sine* and *cosine* rules for finding the sides and angles of non-right-angled triangles, and also three formulae for calculating the areas of triangles.

Figure 10.1 shows a triangle labelled conventionally, and for convenience the formulae are listed below:

The *sine* rule:
$$\frac{a}{\sin A} = \frac{b}{\sin B} = \frac{c}{\sin C}$$

used when given: one side and any two angles,

or: two sides and an angle opposite one of the sides.

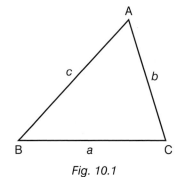

Fig. 10.1

The *cosine* rule:

$$a^2 = b^2 + c^2 - 2bc \cos A$$

or:

$$b^2 = a^2 + c^2 - 2ac \cos B$$

or:

$$c^2 = a^2 + b^2 - 2ab \cos C$$

This is used when given: two sides and the angle between them

or: the three sides.

The area of a triangle may be found using:

either

$$\text{Area} = \tfrac{1}{2} \times \text{base} \times \text{altitude}$$

or

$$\text{Area} = \tfrac{1}{2}ab \sin C = \tfrac{1}{2}bc \sin A = \tfrac{1}{2}ac \sin B$$

or

$$\text{Area} = \sqrt{s(s-a)(s-b)(s-c)} \text{ where } s = \frac{a+b+c}{2}$$

The diameter D of the circumscribing circle of a triangle is given by

$$D = \frac{a}{\sin A} = \frac{b}{\sin B} = \frac{c}{\sin C}$$

EXAMPLE 10.1

Find the resultant of the two forces shown in Fig. 10.2 and the angle it makes with the 50 N force.

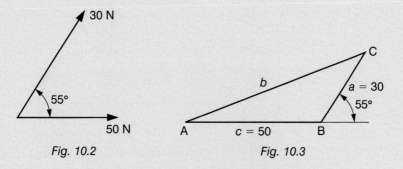

Fig. 10.2   Fig. 10.3

The triangle ABC (Fig. 10.3) is the vector diagram for the given system in which $a = 30$, $c = 50$, and the length b gives the resultant.

Now $\angle ABC = 180° - 55° = 125°$

To find b we will use the cosine rule which gives

$$b^2 = a^2 + c^2 - 2ac \cos B$$

$\therefore$ $$b^2 = 30^2 + 50^2 - 2 \times 30 \times 50 \times \cos 125°$$

$\therefore$ $$b = 71.56$$

To find angle A we will use the sine rule which gives:

$$\frac{a}{\sin A} = \frac{b}{\sin B}$$

from which $$\sin A = \frac{a(\sin B)}{b}$$

$$= \frac{30(\sin 125°)}{71.56}$$

$$A = 20.08°$$

The resultant is 71.56 N and the angle it makes with the 50 N force is 20.08°.

Exercise 10.1

1) The line diagram of a jib crane is shown in Fig. 10.4. Calculate the length of the jib BC and the angle between the jib BC and the tie rod AC.

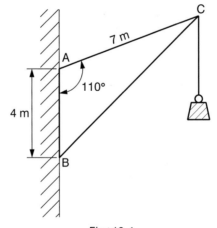

Fig. 10.4

2) The schematic layout of a petrol engine is shown in Fig. 10.5. The crank OC is 80 mm long and the connecting rod CP is 235 mm long. For the position shown find the distance OP and the angle between the crank and the connecting rod.

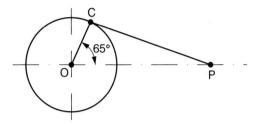

Fig. 10.5

3) A four-bar mechanism ABCD is shown in Fig. 10.6. Links AB, BC and CD are pin-jointed at their ends. (The name 'four-bar' is given because the length AD is considered to be a link.) Find the length AC and the angle ADC.

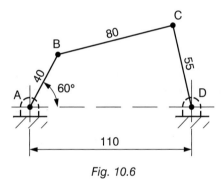

Fig. 10.6

4) Figure 10.7 shows the dimensions between three holes which are to be drilled in a plate of steel. Find the radius of the circle on which the holes lie.

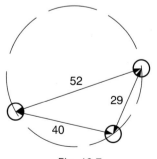

Fig. 10.7

5) P, Q and R are sliders constrained to move along centre-lines OA and OB as shown in Fig. 10.8. For the given position find the distance OR.

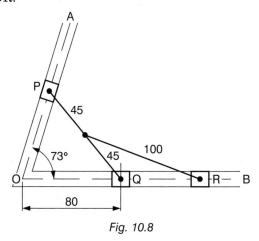

Fig. 10.8

6) Find the lengths of the members BF and CF in the roof truss shown in Fig. 10.9.

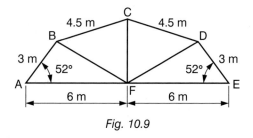

Fig. 10.9

7) Find the area of triangle ABC in Fig. 10.4.

8) Find the area of triangle OCP in Fig. 10.5.

9) Find the area of quadrilateral ABCD in Fig. 10.6.

10) Find the area of the triangle in Fig. 10.7.

Vectors

11

INTRODUCTION

The work we cover on vectors only serves as a basic introduction, as vector analysis is an extensive subject in its own right. An application of vectors is basic navigation using 'velocity triangles'. Here velocities can be represented as vectors, in a similar way to forces. Phasors – special types of vectors – cater for electrical currents and voltages.

SCALARS AND VECTORS

A scalar quantity is one that is fully defined by magnitude alone. Some examples of scalar quantities are time (e.g. 30 seconds), temperature (e.g. 8 degrees Celsius) and mass (e.g. 7 kilograms).

A vector quantity needs magnitude, direction and sense to describe it fully. A vector may be represented by a straight line, its length representing the magnitude of the vector, its direction being that of the vector and suitable notation giving the sense of the vector (Fig. 11.1).

The usual way of naming a vector is to name its end points. The vector in Fig. 11.1 starts at A and ends at B and we write $\overrightarrow{AB}$ which means 'the vector from A to B' – this gives the sense.

Some examples of vector quantities are:

1) A displacement in a given direction, e.g. 15 metres due west.
2) A velocity in a given direction, e.g. 40 km/h due north.
3) A force of 20 kN acting vertically downwards.

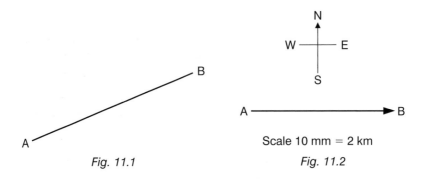

Fig. 11.1   Fig. 11.2

GRAPHICAL REPRESENTATION

A vector may be represented by a straight line drawn to scale.

EXAMPLE 11.1

A man walked a distance of 8 km due east. Draw the vector.

We first choose a suitable scale to represent the magnitude of the vector. In Fig. 11.2 a scale of 10 mm = 2 km has been chosen as convenient. We then draw a horizontal line 40 mm long and label the ends as shown. So $\overrightarrow{AB}$ represents the vector 8 km due east.

Sometimes we add an arrow to the vector (Fig. 11.2) to confirm the sense, but it is not necessary.

CARTESIAN COMPONENTS

Figure 11.3 shows a vector $\overrightarrow{AB}$ of magnitude 6 m/s and making an angle of 30° with the horizontal. An alternative is to define the vector AB by resolving it into its horizontal component $\overrightarrow{AC}$, and its vertical component $\overrightarrow{CB}$. These are said to be the Cartesian components of the vector.

The magnitude of component $\overrightarrow{AC} = 6 \times \cos 30° = 5.20$ m/s

The magnitude of component $\overrightarrow{CB} = 6 \times \sin 30° = 3$ m/s

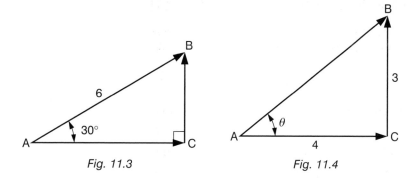

Fig. 11.3   Fig. 11.4

EXAMPLE 11.2

The vector $\overrightarrow{AB}$ in Fig. 11.4 has a vertical component of 3 N and a horizontal component of 4 N. Calculate its magnitude and direction.

Using Pythagoras' theorem the magnitude of $\overrightarrow{AB} = \sqrt{4^2 + 3^3} = 5$ N.

To state its direction we find the size of angle θ.

$$\tan \theta = \tfrac{3}{4} \quad \text{giving} \quad \theta = 36.9°$$

ADDITION OF VECTORS AND RESULTANT

Three points P, Q and R are marked out in a field, and are shown to scale in Fig. 11.5. A man walks from P to Q (i.e. he describes $\overrightarrow{PQ}$), and then walks on from Q to R (i.e. he describes $\overrightarrow{QR}$). Instead the man could have walked directly from P to R, thus describing the vector $\overrightarrow{PR}$.

Now going from P to R directly has the same result as going from P to Q and then from Q to R. We therefore call $\overrightarrow{PR}$ the *resultant* of the sum of the vectors $\overrightarrow{PQ}$ and $\overrightarrow{QR}$,

and we write this as

$$\overrightarrow{PR} = \overrightarrow{PQ} + \overrightarrow{QR}$$

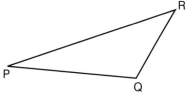

Fig. 11.5

EXAMPLE 11.3

Two forces act at a point O as shown in Fig. 11.6. Find the resultant force at O **a)** by making a scale drawing, **b)** by calculation.

a) The vector diagram is shown in Fig. 11.7. Draw AB parallel to the direction of the 8 N force, and of length representing 8 N to whatever scale you choose. Then draw BC parallel to the direction of the 6 N force and of length representing 6 N to scale.

Here $\overrightarrow{AB} + \overrightarrow{BC} = \overrightarrow{AC}$,

and thus $\overrightarrow{AC}$ represents the resultant of the two given forces. Measure the length AC and the angle it makes with the vertical.

Thus the resultant of the two forces is 11.2 N acting at 13.8° to the vertical.

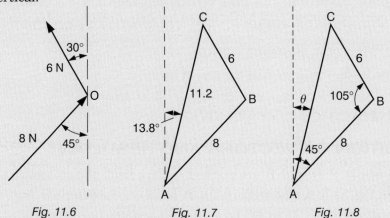

Fig. 11.6   Fig. 11.7   Fig. 11.8

b) Using the cosine rule $b^2 = a^2 + c^2 - 2ac \cos B$
we have from △ABC in Fig. 11.8:

$$AC^2 = 6^2 + 8^2 - 2 \times 6 \times 8 \times \cos 105°$$
$$= 124.8$$

giving $AC = 11.2$ N confirming the drawing result.

Using the sine rule $\dfrac{a}{\sin A} = \dfrac{b}{\sin B}$

or $\sin A = \dfrac{a \sin B}{b}$

We have from △ABC in Fig. 11.8:

$$\sin A = \frac{6 \sin 105°}{11.2}$$

from which $A = \text{inv} \sin 0.5175$
$$= 31.2°$$

Hence $\theta = 45° - 31.2°$
$$= 13.8° \text{ confirming the drawing result.}$$

VECTOR ADDITION USING A PARALLELOGRAM

In the previous example vector addition made use of a triangle.
A parallelogram may also be considered as in Fig. 11.9.

We have $\qquad\qquad\qquad\overrightarrow{OA} + \overrightarrow{OB} = \overrightarrow{OR}$

Using the triangle OAR we would have said that $\overrightarrow{OA} + \overrightarrow{AR} = \overrightarrow{OR}$; this is a
similar statement since $\overrightarrow{OB} = \overrightarrow{AR}$ because they are equal vectors having
the same magnitude, direction and sense. Any numerical calculations
would be carried out using $\triangle OAR$ as in Example 11.3.

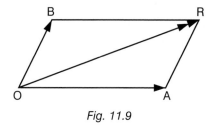
Fig. 11.9

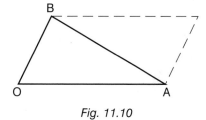
Fig. 11.10

SUBTRACTION OF VECTORS

The inverse of a vector is one having the same magnitude and direction,
but of opposite sense. Thus $\overrightarrow{BA}$ is the inverse of $\overrightarrow{AB}$, or $\overrightarrow{BA} = -\overrightarrow{AB}$.

To subtract a vector we add its inverse. Hence if we wish to subtract
vector $\overrightarrow{OB}$ from $\overrightarrow{OA}$ we have:

$$\overrightarrow{OA} - \overrightarrow{OB} = \overrightarrow{OA} + \overrightarrow{BO} = \overrightarrow{BO} + \overrightarrow{OA} = \overrightarrow{BA}$$

In Fig. 11.10 we can see that from the $\triangle OAB$ a vector sum gives
$$\overrightarrow{BO} + \overrightarrow{OA} = \overrightarrow{BA}.$$

Thus:

Vector difference $\quad \overrightarrow{OA} - \overrightarrow{OB} = \overrightarrow{BA}$

This vector difference is represented by the other diagonal on the
parallelogram than that used for the vector sum.

PHASORS

In electrical engineering currents and voltages are represented by *phasors* in a similar manner to that in which vectors may be used to represent forces and velocities. The methods of adding and subtracting phasors and vectors are similar.

EXAMPLE 11.4

Phasor OA represents 7 V and phasor OB represents 5 V as shown in Fig. 11.11. Find the phasor difference $\overrightarrow{OA} - \overrightarrow{OB}$.

We suggest that you draw a diagram to scale and measure the result – this will serve as a check to the answer obtained by calculation.

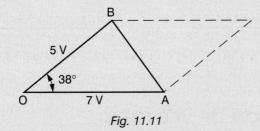

Fig. 11.11

Now $\overrightarrow{OA} - \overrightarrow{OB} = \overrightarrow{BA}$ which confirms that $\overrightarrow{BA}$ is the phasor difference. We are only concerned with lengths and angles in the calculations which follow.

Using the cosine rule for $\triangle OAB$ in Fig. 11.11:

$$AB^2 = OA^2 + OB^2 - 2(OA)(OB) \cos \angle AOB$$
$$= 7^2 + 5^2 - 2 \times 7 \times 5 \times \cos 38°$$
$$= 18.84$$

from which $AB = 4.34$

Thus the phasor difference has a magnitude of 4.34 V and, if required, its direction may be found using the sine rule for $\triangle OAB$.

THE TRIANGLE OF FORCES

This is an important application of vector representation.

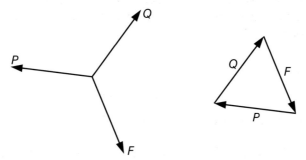

Fig. 11.12   Space diagram   Fig. 11.13   Vector diagram

If three forces act at a point (Fig. 11.12) and are in equilibrium, then they may be represented as vectors (Fig. 11.13) by the sides of a closed triangle.

The word 'closed' refers to the fact that the force vectors follow each other, nose to tail, all the way round the vector diagram with the nose of the last vector finishing up at the tail of the first. This idea may be extended to a polygon vector diagram where each side represents one force of many, in a system which is in equilibrium.

EXAMPLE 11.5

A mass of 7 kg is suspended from the ceiling as in Fig. 11.14. Find the tension forces in the supporting ropes **a)** by scale drawing **b)** by calculation.

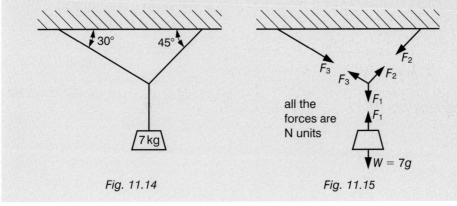

Fig. 11.14   Fig. 11.15

However complicated an engineering system is, it will become much easier if you get into the habit of redrawing the physical arrangement (Fig. 11.15), and splitting it into its component parts. This idea may be extended to frameworks where it simplifies the identification of tension or compression forces in the framework members.

Using either method of solution we shall be considering the equilibrium of the point where the ropes are joined. To find force F_1 we can turn our attention to the 7 kg mass. Since this mass is also in equilibrium then $F_1 = 7g$ N, which is the weight of the mass, and if we take $g = 10 \text{ m s}^{-2}$ then $F_1 = 70$ N.

a) Choose your own scale starting with the known 70 N force vector vertically downwards and draw the other two sides parallel to the lines of action of F_2 and F_3 respectively (Fig. 11.16). Measurement will give you the values of the tension forces $F_2 = 63$ N and $F_3 = 51$ N to 2 s.f. (a reasonable accuracy for a graphical result).

b) Using the sine rule for the force vector triangle:

$$\frac{F_2}{\sin 60°} = \frac{70}{\sin 75°}$$

from which $F_2 = 62.8$ N   correct to 3 s.f.

and using the sine rule again to find F_3:

$$\frac{F_3}{\sin 45°} = \frac{70}{\sin 75°}$$

giving $F_3 = 51.2$ N   correct to 3 s.f.

These results confirm those obtained graphically.

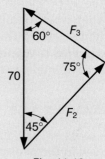

Fig. 11.16

EXAMPLE 11.6

A block of metal is at rest on an incline, as shown in Fig. 11.17. Find:

a) the friction force which is preventing the block from sliding down;

b) the normal reaction between the block and the plane.

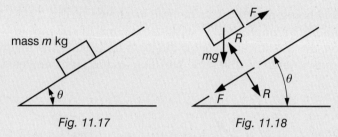

Fig. 11.17   Fig. 11.18

In Fig. 11.18 we show the forces (in newtons) acting on the block itself, and their 'opposites' on the inclined plane (remember Newton's law 'to every action there is an equal and opposite reaction'). Although the forces do not strictly act at a point, the error is minimal and we draw the force vector triangle as in Fig. 11.19.

Since the vector triangle is right-angled then:

$$\sin \theta = \frac{\text{opposite side}}{\text{hypotenuse}}$$

Friction force $F = mg \sin \theta$ newtons

$$\cos \theta = \frac{\text{adjacent side}}{\text{hypotenuse}}$$

Normal reaction $R = mg \cos \theta$ newtons

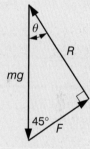

Fig. 11.19

Exercise 11.1

In questions 1–4 find the values of the horizontal and vertical components illustrating your results in each case on a suitable vector diagram.

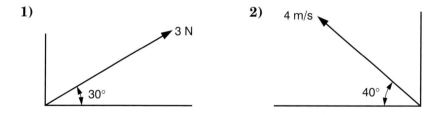

1)

3 N

30°

2)   4 m/s

40°

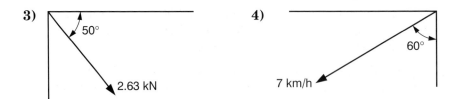

3) 50°   2.63 kN

4) 60°   7 km/h

5) A block is being pulled up an incline, as shown in Fig. 11.20, using a rope making an angle of 25° with the incline. If the pull in the rope is 50 kN, draw a suitable vector diagram indicating the magnitudes of the components of this force parallel, and at right angles, to the incline.

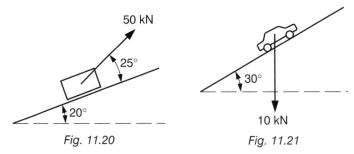

Fig. 11.20   Fig. 11.21

6) A motor car weighing 10 kN is shown on a 30° slope in Fig. 11.21. Resolve this weight into components along the slope and at right angles to it.

7) A garden roller is being pushed forwards with a force of 95 N as shown in Fig. 11.22. Find the horizontal and vertical components of this force.

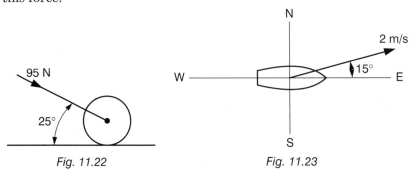

Fig. 11.22   Fig. 11.23

8) A ship is being steered in an easterly direction as shown in plan view in Fig. 11.23. However, owing to a S to N crosswind its motion and velocity are as shown. Find the forward speed of the vessel.

9) A sledge is being pulled on the level by two horizontal ropes which are at right angles to each other. If the respective tensions in the ropes are 30 N and 50 N, find the resultant force on the sledge and the angle it makes with the rope having the greater tension.

10) A plane is flying due north with a speed of 700 km/h. If there is a west to east cross-wind of 70 km/h, determine the resultant velocity of the plane and give its direction relative to north.

11) A dinghy is drifting under the influence of a 0.5 m/s tide flowing from north-east to south-west, and a wind of 0.8 m/s blowing from south to north. Find the resultant velocity of the dinghy and its direction relative to north.

12) The plan of a truck on rails is shown in Fig. 11.24. Find the resultant pull of the two ropes and the direction it acts in relative to the line of the rails.

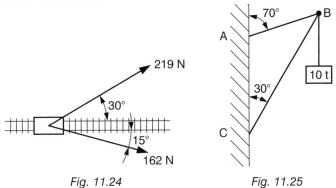

Fig. 11.24   Fig. 11.25

13) Using the results obtained to question **12** find the force component pulling the truck along the track, and the force between the wheel flanges and the rails.

In questions **14–17** find the resultants of the given phasors: in each case giving the direction relative to the larger given value.

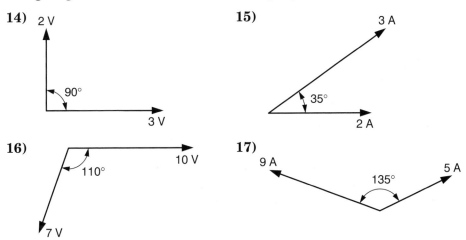

18) Find the forces in the members of the jib crane shown in Fig. 11.25. The units should be kN and you should take care to indicate if the member is in tension or compression.

12 Trigonometric Waveforms

Radian measure – approximations for small angles – amplitude or peak value – angular and time base – cycle – period – frequency – phasors – phase angle – phasor addition – combining sine waves of the same and of different frequencies – harmonics – Fourier series.

INTRODUCTION

Remember radians ? – the alternative way of measuring an angle rather than degrees – good. Then we proceed to look again at the properties of the more common waveforms – sine and cosine curves.

Often two waveforms are present and it may be necessary to combine these and obtain one new curve.

Finally a brief introduction to harmonic analysis – sounds complicated but you maybe pleasantly surprised and find it fascinating, especially as there are free programs on the internet that allow you to play around with waveforms without any mathematical calculations !!!

RADIAN MEASURE

We have seen that an angle is usually measured in degrees but there is another way of measuring an angle. In this, the unit is known as the radian (abbreviation rad).

Referring to Fig. 12.1 gives:

$$\text{Angle in radians} = \frac{\text{length of arc}}{\text{radius of circle}}$$

$$\theta \text{ radians} = \frac{l}{r}$$

$$l = r\theta$$

Hence   Length of arc $= r\theta$

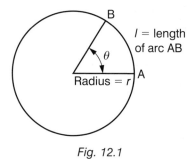

Fig. 12.1

RELATION BETWEEN RADIANS AND DEGREES

For a full $360°$ circle the arc will be the circumference $2\pi r$

thus   Full circle angle $= \dfrac{\text{circumference}}{\text{radius}} = \dfrac{2\pi r}{r} = 2\pi$ radians

giving $360° = 2\pi$ radians

so 1 radian $= \dfrac{360°}{2\pi} = 57.3$ degrees

To convert from degrees to radians	To convert from radians to degrees
angle $= \dfrac{\pi(\theta°)}{180}$ radians	angle $= \left(\dfrac{180}{\pi \times \theta}\right)$ degrees

More often than not we use ratios of π for angles measured in radians, rather than using a numerical value. Here are the most used angles:

$30° = \dfrac{\pi(30)}{180} = \dfrac{\pi}{6}$ rad		$45° = \dfrac{\pi}{4}$ rad	$60° = \dfrac{\pi}{3}$ rad
$90° = \dfrac{\pi}{2}$ rad	$180° = \pi$ rad	$270° = \dfrac{3\pi}{2}$ rad	$360° = 2\pi$ rad

DEGREES, MINUTES AND SECONDS

Modern calculating methods make the use of decimal degrees (e.g. $36.783°$) more likely than the use of minutes and seconds.

Most scientific calculators will convert degrees, minutes and seconds into decimal degrees and vice versa, using special keys – instructions for use of these keys will be given in the accompanying booklet for the calculator.

EXAMPLE 12.1

Convert $29°\,37'\,29''$ to radians stating the answer correct to 5 significant figures.

The first step is to convert the given angle into degrees and decimals of a degree.

$$29°\,37'\,29'' = 29 + \dfrac{37}{60} + \dfrac{29}{3600} = 29.625° = \dfrac{\pi \times 29.625}{180} = 0.517\,05 \text{ radians}$$

EXAMPLE 12.2

Convert 0.089 35 radians into degrees.

$$0.089\,35 \text{ radians} = \frac{0.089\,35 \times 180}{\pi} = 5.119°$$

Approximations for Small Angles

If the angle θ is small, i.e. less than about 0.1 radians or 6°, the following approximations are often used:

$\sin \theta \approx \theta$	
$\cos \theta \approx 1 - \dfrac{\theta^2}{2}$	Providing θ is in *radians*
$\tan \theta \approx \theta$	

For example, when $\theta = 0.1$ rad (5.73°):

Function	Calculator value to 4 s.f.	Approximate value
$\sin 0.1$	0.0998	0.1000
$\cos 0.1$	0.9950	$1 - \dfrac{(0.1)^2}{2} = 0.9950$
$\tan 0.1$	0.1003	0.1000

AMPLITUDE OR PEAK VALUE

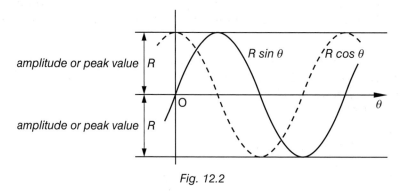

Fig. 12.2

In Fig. 12.2 the value **2R** is called the **double amplitude** or, more commonly, the **peak to peak value** for obvious reasons.

Graphs of $\sin\theta$, $\sin 2\theta$, $2\sin\theta$ and $\sin\frac{1}{2}\theta$

Curves of the above trigonometrical functions are shown plotted in Fig. 12.3. You may find it useful to construct the curves using values obtained from a calculator.

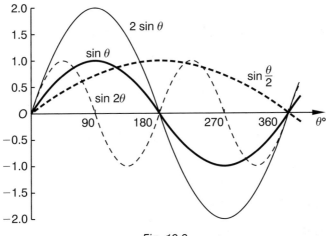

Fig. 12.3

Graphs of $\cos\theta$, $\cos 2\theta$, $2\cos\theta$ and $\cos\frac{1}{2}\theta$

Cosine graphs are similar in shape to sine curves. You should plot graphs of the above functions from $0°$ to $360°$ using values obtained from a calculator.

Graphs of $\sin^2\theta$ and $\cos^2\theta$

It is sometimes necessary in engineering applications, such as when finding the root mean square value of alternating currents and voltages, to be familiar with the curves $\sin^2\theta$ and $\cos^2\theta$.

Values of the functions can be obtained using a calculator and their graphs are shown in Fig. 12.4. We should note that the curves are wholly positive, since squares of negative or positive values are always positive.

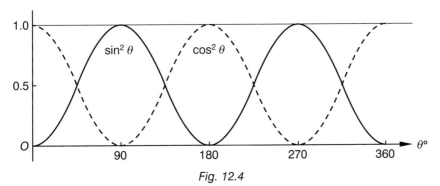

Fig. 12.4

Relation between Angular and Time Scales

In Fig. 12.5 OP represents a radius, of length R, which rotates at a uniform angular velocity ω radians per second about O, the direction of rotation being anticlockwise.

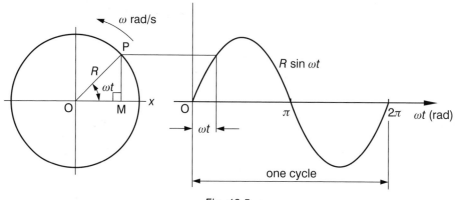

Fig. 12.5

$$\text{Angular velocity} = \frac{\text{Angle turned through}}{\text{Time taken}}$$

$\therefore$ Angle turned through $=$ (Angular velocity) $\times$ (Time taken)

$$= \omega t$$

Also $PM \backslash OP = \sin \hat{POM}$ from right-angled $\triangle OPM$

$\therefore$ $PM = OP \sin \hat{POM}$

or $PM = R \sin \omega t$

If a graph is drawn, as in Fig. 12.5, showing how PM varies with the angle ωt, the sine wave representing $R \sin \omega t$ is obtained. It can be seen that the **peak value** of this sine wave is R (i.e. the magnitude of the rotating radius).

The horizontal scale shows the angle turned through, ωt, and the waveform is said to be plotted on an **angular** or ωt **base**.

Cycle

A **cycle** is the portion of the waveform which shows its **complete shape without any repetition**. It may be seen from Fig. 12.5 that one cycle is completed whilst the radius OP turns through 360° or 2π radians.

Period

This is the **time** taken for the waveform to **complete one cycle**. It will also be the time taken for OP to complete one revolution or 2π radians.

Now we know that   Time taken $= \dfrac{\text{Angle turned through}}{\text{Angular velocity}}$

Hence

$$\text{The period} = \frac{2\pi}{\omega} \text{ seconds}$$

Frequency

The number of **cycles per second** is called the **frequency**. The unit of frequency representing one cycle per second is the hertz (Hz).

Now if   1 cycle is completed in $\dfrac{2\pi}{\omega}$ seconds (**a period**)

then $1 \div \dfrac{2\pi}{\omega}$ cycles are completed in 1 second

or $\dfrac{\omega}{2\pi}$ cycles are completed in 1 second

Hence $$\text{frequency} = \frac{\omega}{2\pi} \text{ Hz}$$

and since period $= \dfrac{2\pi}{\omega}$ then $$\text{frequency} = \frac{1}{\text{period}}$$

Graphs of $\sin t$, $\sin 2t$, $2\sin t$ and $\sin\frac{1}{2}t$

Now waveform $\sin \omega t$ has a period of $\dfrac{2\pi}{\omega}$ seconds.

Thus waveform $\sin t$ has a period of $\dfrac{2\pi}{1} = 6.26$ seconds.

We have seen how a graph may be plotted on an 'angular' or ωt' base as in Fig. 12.5. Alternatively, the **units on the horizontal axis** may be those of **time (usually seconds)**, and this is called a **'time' base**, as displayed on an oscilloscope.

In order to plot one complete cycle of the waveform it is necessary to take values of t from 0 to 6.28 seconds. We suggest that you plot the curve of $\sin t$, remembering to set your calculator to the 'radian' mode when finding the value of $\sin t$. The curve is shown plotted on a time base in Fig. 12.6.

Similarly,

$$\text{The waveform } \sin 2t \text{ has a period of } \frac{2\pi}{2} = 3.14 \text{ seconds}$$

$$\text{and waveform } \sin \tfrac{1}{2}t \text{ has a period of } \frac{2\pi}{\frac{1}{2}} = 12.56 \text{ seconds}$$

Each of the above waveforms has an amplitude of unity. However, the waveform $2\sin t$ has an amplitude of 2, although its period is the same as that of $\sin t$, namely 6.28 seconds.

All these curves are shown plotted in Fig. 12.6. This enables a visual comparison to be made and it may be seen, for example, that the curve of $\sin 2t$ has a frequency twice that of $\sin t$ (since two cycles of $\sin 2t$ are completed during one cycle of $\sin t$).

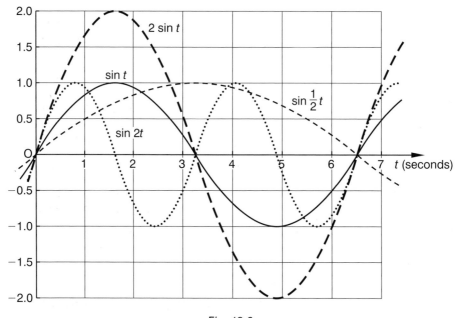

Fig. 12.6

Graphs of $R\cos \omega t$

The waveforms represented by $R \cos \omega t$ are similar to sine waveforms, R being the peak value and $\dfrac{2\pi}{\omega}$ the period. You are left to plot these as instructed in the following exercise.

Exercise 12.1

1) On the same axes, using a time base, plot the waveforms of $\cos t$ and $2 \cos t$ for one complete cycle from $t = 0$ to $t = 6.28$ seconds.

2) Using the same axes on which the curves were plotted in question 1, plot the waveforms of $\cos 2t$ and $\cos \dfrac{t}{2}$.

3) On the same axes, using an angle base from $0°$ to $360°$, sketch the following waveforms:

a) $5 \cos \theta$ **b)** $3 \sin 2\theta$ **c)** $4 \cos 3\theta$ **d)** $2 \sin 3\theta$.

PHASE ANGLE

The principal use of sine and cosine waveforms occurs in electrical technology in which they represent alternating currents and voltages. In a diagram such as that shown in Fig. 12.7 the rotating radii OP and OQ are called **phasors**.

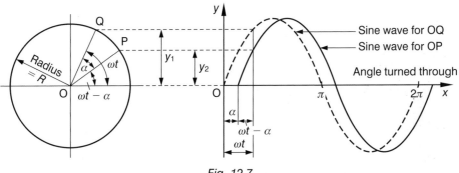

Fig. 12.7

Figure 12.7 shows two phasors OP and OQ, separated by an angle a, rotating at the same angular speed in an anticlockwise direction. The sine waves produced by OP and OQ are identical curves but they are displaced from each other. The amount of displacement is known as the **phase difference** and, measured along the horizontal axis, is a. The angle a is called the **phase angle**.

In Fig. 12.7 the phasor OP is said to **lag** behind phasor OQ by the angle a. If the radius of the phasor circle is R then OP $=$ OQ $= R$ and hence

for the phasor OQ, $y_1 = R \sin \omega t$

and for the phasor OP, $y_2 = R \sin(\omega t - a)$

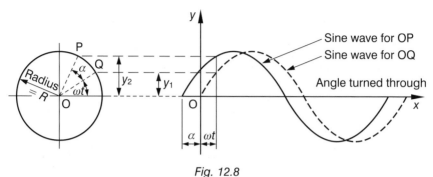

Fig. 12.8

Similarly in Fig. 12.8 the phasor OP **leads** the phasor OQ by the phase angle a.

Hence for the phasor OQ

$$y_1 = R \sin \omega t$$

and for the phasor OP

$$y_2 = R \sin(\omega t + a)$$

In practice it is usual to draw waveform on an 'angular' or 'ωt' base when considering phase angles, as in the following example.

EXAMPLE 12.3

Sketch the waveforms of $\sin \omega t$ and $\sin\left(\omega t - \dfrac{\pi}{3}\right)$ on an angular base and identify the phase angle.

The curve $\sin \omega t$ will be plotted between $\omega t = 0$ and $\omega t = 2\pi$ radians (i.e. over 1 cycle).

Also $\sin\left(\omega t - \dfrac{\pi}{3}\right)$ will be plotted between values given by

$$\omega t - \frac{\pi}{3} = 0 \quad \text{and when} \quad \omega t - \frac{\pi}{3} = 2\pi$$

i.e. $\omega t = \dfrac{\pi}{3}$ radians   and $\omega t = 2\pi + \dfrac{\pi}{3} = \dfrac{7\pi}{3}$ radians

The graphs are shown in Fig. 12.9.

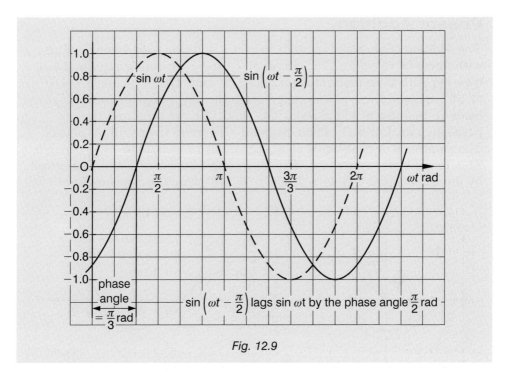

Fig. 12.9

COMBINING SINE WAVES

In the context of the work which follows 'combining' sine waves means *adding sine waves together*. The result of this addition is called the **resultant waveform**. We shall examine the result of adding two sine waves of the same frequency and also of adding two sine waves of different frequencies.

Adding Two Sine Waves of the Same Frequency

The methods used are well illustrated by means of a typical example.

Consider two sinusoidal electric currents represented by the equations $i_1 = 10 \sin \theta°$ and $i_2 = 12 \sin(\theta + 30)°$. If the resultant waveform is denoted by i_r, then

$$i_r = i_1 + i_2$$
$$= 10 \sin \theta° + 12 \sin(\theta + 30)°$$

The addition may be achieved by either phasor addition or addition of the sine waves.

Phasor Addition

The addition of phasors is similar to vector addition used when dealing with forces or velocities. The curves of i_1 and i_2 are shown in Fig. 12.10 together with their associated phasors OC' and OD'. The angle $C'OD'$ of $30°$ between these phasors represents the phase angle by which i_2 leads i_1. In order to add phasors OC' and OD' we construct the parallelogram $OC'B'D'$. Then diagonal OB' gives the resultant phasor. The amplitude or peak value of i_r is given by the length of OB' which is 21.3, found by measurement. The phase angle between i_r and i_1 is given by the angle $C'OB'$. Thus i_r leads i_1 by $16°$, found by measurement.

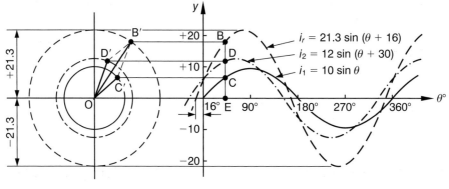

Fig. 12.10

Addition of the Sine Waves

In order to plot the waveform for i_r we need values of the expression $10 \sin \theta° + 12 \sin(\theta + 30)°$ for suitable values of θ from $0°$ to $360°$. A suitable calculator sequence using, for example, $\theta = 25°$ is as follows, remembering to set first the angle mode on the machine to 'degrees'. Recall that a calculator will 'do' $\sin 25$ before adding 30 if we don't use brackets, and noting it is possible to omit some $\times$ signs e.g. between 10 and sin, a suitable sequence is:

$$\boxed{AC}\ \boxed{10}\ \boxed{\sin}\ \boxed{25}\ \boxed{+}\ \boxed{12}\ \boxed{\sin}\ \boxed{(}\ \boxed{25}\ \boxed{+}\ \boxed{30}\ \boxed{)}\ \boxed{=}$$

giving an answer 14.1 correct to three significant figures.

The graphs of i_1, i_2 and i_r are shown in Fig. 12.10 where it will be seen that i_r is a sine wave with a peak value of 21.3 and a phase angle of $16°$. It has the equation $i_r = 21.3 \sin(\theta + 16°)$.

The peak value of 21.3 and the phase angle of $16°$ are only approximate as their accuracy depends on reading values from the scales of the graphs. In this case it is possible to calculate these values by a theoretical method (beyond the scope of this book) and obtain more accurate answers of 21.254 and $16°24'$ which shows the answers we obtained are as good as could be expected from a graphical method.

It is possible to obtain the i_r curve from the graphs of i_1 and i_2 by graphical addition. This is, in fact, similar to adding the values of i_1 and i_2 for various values of the angle θ.

First plot the graphs of i_1 and i_2 and then the points on the i_r curve may be plotted by adding the ordinates (i.e. the vertical lengths) of the corresponding points on the i_1 and i_2 curves.

For example, to plot point B on the i_r curve in Fig. 12.10 we can measure the lengths CE and DE and then add their values, since BE = CE + DE.

We must take care to allow for the ordinates being positive or negative.

Adding Two Sine Waves of Different Frequencies

Since we desire to see the shape of the resultant curve, we shall add the given sine waves. Let us see the result of adding together the sine curves $\sin \theta$ and $\sin 2\theta$. Note that $\sin 2\theta$ has a frequency twice that of $\sin \theta$.

In order to plot the resultant curve we need values of the expression $(\sin \theta + \sin 2\theta)$. As usual we shall use values of θ from $0°$ to $360°$.

The curves of $\sin \theta$, $\sin 2\theta$ and $(\sin \theta + \sin 2\theta)$ are shown plotted in Fig. 12.11.

We can see that the graph of $(\sin \theta + \sin 2\theta)$ is non-sinusoidal (i.e. *not* the shape of a sine curve). However, the curve is a waveform (or a periodic function) since it will repeat indefinitely.

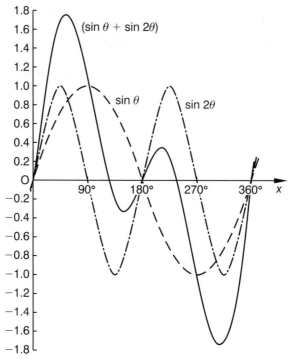

Fig. 12.11

In general the addition of any two sine waves of different frequencies will result in a non-sinusoidal waveform. You may check this by working through the examples in question 10 of Exercise 12.2 which follows.

Exercise 12.2

1) Plot the graphs of $\sin \theta$ and $\sin(\theta + 0.9)$ on the same axes on an angular base using units in radians. Indicate the phase angle between the waveforms and explain whether it is an angle of lead or lag.

2) Plot the graph of $\sin\left(\omega t + \dfrac{\pi}{6}\right)$ and $\sin \omega t$ on the same axes on an angular base showing a cycle of each waveform. Identify the phase angle between the curves.

3) Plot the graphs of $\sin\left(\omega t + \dfrac{\pi}{3}\right)$ and $\sin\left(\omega t - \dfrac{\pi}{4}\right)$ on the same axes on an angular base showing a cycle of each waveform. Identify the phase angle between the curves.

4) Write down the equation of the waveform which:

   a) leads $\sin \omega t$ by $\dfrac{\pi}{2}$ radians.

   b) lags $\sin \omega t$ by π radians.

   c) leads $\sin\left(\omega t - \dfrac{\pi}{3}\right)$ by $\dfrac{\pi}{3}$ radians

   d) lags $\sin\left(\omega t + \dfrac{\pi}{6}\right)$ by $\dfrac{\pi}{3}$ radians

5) Plot the waves of $\sin \theta$ and $\cos \theta$ on the same axes on an angular base showing a cycle of each waveform. Identify the phase angle between the curves.

6) Find the resultant waveform of the curves $3 \sin \theta$ and $2 \cos \theta$ by graphical addition. What is the amplitude of the resultant waveform and the phase angle relative to $3 \sin \theta$?

7) Find the resultant voltage v_r of the two voltages represented by the equation $v_1 = 3 \sin \theta$ and $v_2 = 5 \sin(\theta - 30°)$ by plotting the three graphs.

8) Plot the graphs of $i_1 = 5 \sin \theta$ and $i_2 = 2 \sin(\theta + 45°)$ and hence find the resultant of i_1 and i_2 by graphical addition. State the equation of the resultant current i_r.

9) The voltages v_1 and v_2 are represented by the equations

$$v_1 = 30 \sin(\theta + 60°) \quad \text{and} \quad v_2 = 50 \sin(\theta - 45°).$$

Plot the curves of v_1 and v_2 and the resultant voltage v_r and find the equation representing v_r.

10) Find the resultant curve in each of the following and verify that the resultant is a non-sinusoidal periodic waveform:

   a) $\sin \theta + \sin \dfrac{\theta}{2}$ **b)** $3 \sin \theta + 2 \sin 2\theta$

   c) $\sin \dfrac{\theta}{2} + 2 \sin 2\theta$ **d)** $\sin(\theta + 30°) + \sin 2\theta$

HARMONICS

Let us start with a waveform which has, what we deem to be, the basic fundamental frequency (Fig. 12.12).

Then harmonics are frequencies which are integer multiples of the fundamental.

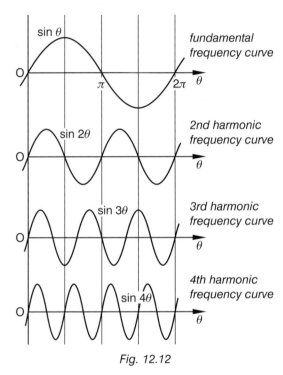

Fig. 12.12

Those waveforms shown have approximately the same amplitudes, but usually as the harmonic increases then its amplitude gets less.

FOURIER SERIES

This comprises the sum of an infinite number of sine functions. The first term has a chosen fundamental frequency whilst the rest are harmonics:

$$a_1 \sin \theta \quad + \quad a_2 \sin 2\theta \quad + \quad a_3 \sin 3\theta \quad + \quad a_4 \sin 4\theta \quad + \dots\dots\dots$$

The coefficients a_1, a_2, a_3, ...etc are the amplitudes of each harmonic waveform.

If we are clever enough to choose 'suitable' values for these coefficients and then add the curves as previously in this chapter we will obtain a periodic waveform. This will have the fundamental frequency and its shape will depend on the values we pick for the amplitude coefficients – they may be positive, negative or zero.

The more common of our aims are triangular, sawtooth (or ramp), square, and rectified waveforms.

Since this is only an introduction we will use a **square waveform** to explain how the system works.

A good thing is that we only have to examine, in turn, the first few terms to get a good idea of how to proceed.

Figure 12.13 shows a square waveform and Fig. 12.14 shows our fundamental sine curve to a suitable size.

Since the amplitude of the square wave is 1 then we will let $a_1 = 1$.

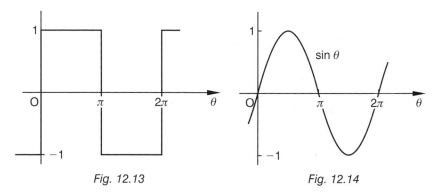

Fig. 12.13   Fig. 12.14

Imagine that the portion of the sine wave shown above the axis is an elastic cord secured at O and π. We then grasp it about two thirds up each side and pull it to a square shape like the waveform. Obviously we can't do this mathematically but we achieve the same result by adding and subtracting ordinates so that its shape will change to a square waveform – sounds impossible but let us see what happens.

If we start by including the second harmonic we will have a curve whose equation is $a_1 \sin \theta + a_2 \sin 2\theta$.
Now we have already some experience of this when we plotted
$\sin \theta + \sin 2\theta$ and even if we changed the amplitude coefficients the result would still be 'lop-sided' and not what we want. So let's discard this term by making $a_2 = 0$.

Moving on to the third harmonic we note that it is symmetrical over the length from O to π and by experimenting with values we find that $a_3 = \dfrac{1}{3}$ gives the result shown in Fig. 12.15 – at least we can see that the shape is changing towards our goal.

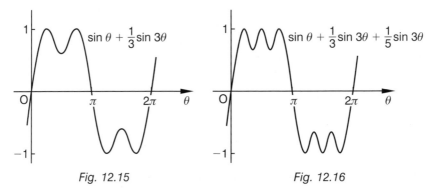

Fig. 12.15   Fig. 12.16

The fourth harmonic proves as ineffective as the second so we let $a_4 = 0$

The fifth harmonic is also useful so we again add using $a_5 = \dfrac{1}{5}$ and this result is shown in Fig. 12.16.

Now we could go on following the pattern and a very near approximation will be achieved as more terms are included.

FOURIER SERIES ON THE INTERNET – FREE AND NO CALCULATIONS!

If you have access to the internet there are a number of programs available free. Go to a search engine such as 'google' and type in 'fourier series' and a number of items will appear. If you access some of these you will find that it is possible to 'build' a waveform using values of the coefficients set using scroll bars. This enables you to get some feeling for the effect of changing coefficient values and understand that nearly all periodic functions may be represented by Fourier series – so you can learn all about Fourier series without doing any calculations at all !!!

Differentiation **13**

Gradient of a curve – differentiation of a sum of terms which may include
ax^n, $\sin ax$, $\cos ax$, $\log_e x$, e^{ax} – differentiation by substitution –
differentiation of products and quotients.

INTRODUCTION

Here we shall revise the basic ideas of differential calculus. Do you
remember how we measured the gradient of graphs and found it easier to
use a theoretical way to do the same – a pity the name was designed to put
you off! Anyway, we shall extend the scope of the expressions we can
handle, and this will lead us to some practical technology applications in
the next chapter.

GRADIENT OF A TANGENT

You may recall that the process of differentiation is a method of finding
the rate of change of a function. The rate of change at a point on a curve
may be found by determining the gradient of the tangent at that point.
This may be achieved by either a theoretical or graphical method as
follows.

Suppose we wish to find the gradient of the curve $y = x^2$ at the points
where $x = 3$ and $x = -2$.

From a theoretical approach you may remember that if the function is of
the form $y = x^n$, then the process of differentiating gives $\dfrac{dy}{dx} = nx^{n-1}$.

Hence if
$$y = x^2$$

then
$$\frac{dy}{dx} = 2x$$

and therefore when $x = 3$, the gradient of the tangent

$$\frac{dy}{dx} = 2(3) = +6$$

and when $x = -2$, the gradient of the tangent

$$\frac{dy}{dx} = 2(-2) = -4$$

Alternatively is we decide to use the graphical method we must first draw the graph as shown in Fig. 13.1.

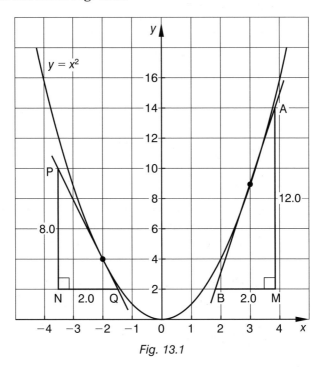

Fig. 13.1

Tangents have been drawn at the given points, that is where $x = 3$ and $x = -2$, and then the gradients may be found by constructing suitable right-angled triangles.

Where $\quad x = 3 \quad$ the gradient $= \dfrac{\text{AM}}{\text{BM}} = \dfrac{12.0}{2.0} = +6$

and where $\quad x = -2 \quad$ the gradient $= \dfrac{\text{PN}}{\text{QN}} = -\dfrac{8.0}{2.0} = -4$

These results verify those obtained by the theoretical method.

DIFFERENTIATION OF A SUM

To differentiate an expression containing the sum of several terms, we differentiate each individual term separately.

Hence if $\qquad y = x^4 + 2x^3 + 5x^2 + 7$

then $\qquad \dfrac{dy}{dx} = 4x^3 + 6x^2 + 10x$

And if $\qquad y = ax^3 + bx^2 - cx + d$

then $\qquad \dfrac{dy}{dx} = 3ax^2 + 2bx - c$

And if $\qquad y = \sqrt{x} + \dfrac{1}{\sqrt{x}} = x^{1/2} + x^{-1/2}$

then $\qquad \dfrac{dy}{dx} = \frac{1}{2}x^{-1/2} + (-\frac{1}{2})x^{-3/2} = \dfrac{1}{2\sqrt{x}} - \dfrac{1}{2\sqrt{x^3}}$

And if $\qquad y = 3.1x^{1.4} - \dfrac{3}{x} + 5 = 3.1x^{1.4} - 3x^{-1} + 5$

then $\qquad \dfrac{dy}{dx} = (3.1)(1.4)x^{0.4} - 3(-1)x^{-2} = 4.34x^{0.4} + \dfrac{3}{x^2}$

And if $\qquad y = \dfrac{t^3 + t}{t^2} = \dfrac{t^3}{t^2} + \dfrac{t}{t^2} = t + t^{-1}$

then $\qquad \dfrac{dy}{dt} = 1 + (-1)t^{-2} = 1 - \dfrac{1}{t^2}$

Exercise 13.1

1) Find $\dfrac{dy}{dx}$ if $y = 5x^3 + 7x^2 - x - 1$

2) If $s = 7\sqrt{t} - 6t^{0.3}$, find an expression for $\dfrac{ds}{dt}$

3) Find by a theoretical method the value of $\dfrac{dy}{dx}$ when $x = 2$ for the curve $y = x - \dfrac{1}{x}$

4) Find the expression for $\dfrac{dy}{du}$ if $y = \dfrac{u + u^2}{u}$

5) Find graphically the gradient of the curve $y = x^2 + x + 2$ at the points where $x = +2$ and $x = -2$. Check the result by differentiation.

DIFFERENTIATION BY SUBSTITUTION

Up to now we have only learnt how to differentiate comparatively simple expressions such as $y = 3x^{4.5}$. For more difficult expressions such as $y = \sqrt{(x^3 + 3x - 9)}$ we make a substitution.

If we put $u = x^3 + 3x - 9$ then $y = \sqrt{u}$.

Now y is a function of u, and since u is a function of x, it follows that y is a function of a function of x.

This all sounds rather complicated and the words 'function of a function' should be noted in case you meet them again, but we prefer to use the expression 'differentiation by substitution'.

A substitution method is often used for differentiating the more complicated expressions, together with the formula

$$\frac{dy}{dx} = \frac{dy}{du} \times \frac{du}{dx}$$

EXAMPLE 13.1

Find $\dfrac{dy}{dx}$ if $y = (x^2 - x)^9$

We have $\qquad\qquad y = (x^2 - x)^9$

Then $\qquad\qquad y = u^9 \qquad$ where $\quad u = x^2 - x$

$\therefore \qquad\qquad \dfrac{dy}{du} = 9u^8 \qquad$ and $\quad \dfrac{du}{dx} = 2x - 1$

But $\qquad\qquad \dfrac{dy}{dx} = \dfrac{dy}{du} \times \dfrac{du}{dx}$

$\therefore \qquad\qquad \dfrac{dy}{dx} = 9u^8 \times (2x - 1)$

The differentiation has now been completed and it only remains to put u in terms of x by using our original substitution $u = x^2 - x$.

Hence $\qquad\qquad \dfrac{dy}{dx} = 9(x^2 - x)^8(2x - 1)$

EXAMPLE 13.2

Find $\dfrac{d}{dx}(\sqrt{(1-5x^3)})$

$\dfrac{d}{dx}(\sqrt{(1-5x^3)})$ is called the differential coefficient of $\sqrt{(1-5x^3)}$ with respect to x. This simply means that we have to differentiate the expression with respect to x. If we let $y = \sqrt{(1-5x^3)}$, then the problem is to find $\dfrac{dy}{dx}$.

Let	$y = \sqrt{(1-5x^3)}$	
i.e.	$y = (1-5x^3)^{1/2}$	
Then	$y = u^{1/2}$	where $u = 1-5x^3$

$\therefore \qquad \dfrac{dy}{du} = \tfrac{1}{2}u^{-1/2}$ and $\dfrac{du}{dx} = -15x^2$

But $\qquad \dfrac{dy}{dx} = \dfrac{dy}{du} \times \dfrac{du}{dx}$

$\therefore \qquad \dfrac{dy}{dx} = \tfrac{1}{2}u^{-1/2} \times (-15x^2)$

$\therefore \qquad \dfrac{dy}{dx} = \tfrac{1}{2}(1-5x^3)^{-1/2}(-15x^2)$

$$= -\frac{15}{2}x^2(1-5x^3)^{-1/2}$$

$$= -\frac{15x^2}{2(\sqrt{1-5x^3})}$$

Hence

$$\frac{d}{dx}(\sqrt{(1-5x^3)}) = -\frac{15x^2}{2\sqrt{(1-5x^3)}}$$

DIFFERENTIATION BY RECOGNITION

Consider $y = (\ )^n$ where any function of x can be written inside the bracket. Then differentiating with respect to x, we have

$$\frac{dy}{dx} = \frac{dy}{d(\ )} \times \frac{d(\ )}{dx}$$

Thus to differentiate an expression of the type $(\ )^n$, first differentiate the bracket, treating it as a term similar to x^n. Then differentiate the function x inside the bracket. Finally, to obtain an expression for $\frac{dy}{dx}$, multiply these two results together.

EXAMPLE 13.3

Find $\frac{dy}{dx}$ if $y = (x^2 - 5x + 3)^5$

Differentiating the bracket as a whole we have

$$\frac{dy}{d(\ )} = 5(x^2 - 5x + 3)^4 \qquad [1]$$

Also the function inside the bracket is $x^2 - 5x + 3$.

Differentiating this gives

$$\frac{d(\ )}{dx} = 2x - 5 \qquad [2]$$

Thus multiplying the results [1] and [2] together gives

$$\frac{dy}{dx} = 5(x^2 - 5x + 3)^4 \times (2x - 5)$$

$$= 5(2x - 5)(x^2 - 5x + 3)^4$$

Hence by recognising the method we can differentiate directly.
If, for example, $y = (x^2 - 3x)^7$

then $$\frac{dy}{dx} = 7(x^2 - 3x)^6 \times (2x - 3)$$

$$= 7(2x - 3)(x^2 - 3x)^6$$

DIFFERENTIATION OF TRIGONOMETRIC FUNCTIONS

It can be shown that, for the standard forms of trigonometric expressions, where ω and α are constants:

$$\frac{d}{dt}\sin(\omega t + \alpha) = \omega\cos(\omega t + \alpha)$$

$$\frac{d}{dt}\cos(\omega t + \alpha) = -\omega\sin(\omega t + \alpha)$$

EXAMPLE 13.4

Find $\frac{dy}{dt}$ if $y = \cos\left(2t + \frac{\pi}{2}\right)$

If we compare this with the standard form, then $\omega = 2$ and $\alpha = \frac{\pi}{2}$

Thus if
$$y = \cos\left(2t + \frac{\pi}{2}\right)$$

then
$$\frac{dy}{dt} = -2\sin\left(2t + \frac{\pi}{2}\right)$$

DIFFERENTIATION OF LOGARITHMIC FUNCTIONS

It can be shown that:

$$\frac{d}{dx}(\log_e x) = \frac{1}{x}$$

EXAMPLE 13.5

Find $\frac{d}{dx}(\log_e 2x)$

Let
$$\begin{aligned}
y &= \log_e 2x \\
&= \log_e 2 + \log_e x \quad \text{using law 1 of logs for} \\
&\qquad\qquad\qquad\qquad \text{numbers multiplied}
\end{aligned}$$

Now $\log_e 2$ is simply a constant, so it will disappear on differentiation

Thus if
$$y = \log_e 2 + \log_e x$$

then
$$\frac{dy}{dx} = \frac{1}{x}$$

DIFFERENTIATION OF EXPONENT FUNCTIONS

It can be shown, where a is a constant, that

$$\frac{\mathrm{d}}{\mathrm{d}x}(\mathrm{e}^{ax}) = a\mathrm{e}^{ax}$$

EXAMPLE 13.6

Find $\dfrac{\mathrm{d}}{\mathrm{d}x}(\mathrm{e}^{5t} + \mathrm{e}^{-2t})$

As usual, when dealing with terms added together, we deal with each term individually.

Thus if we let $\qquad y = \mathrm{e}^{5t} + \mathrm{e}^{-2t}$

then $\qquad\qquad \dfrac{\mathrm{d}y}{\mathrm{d}x} = 5\mathrm{e}^{5t} + (-2)\mathrm{e}^{-2t} = 5\mathrm{e}^{5t} - 2\mathrm{e}^{-2t}$

EXAMPLE 13.7

Find $\dfrac{\mathrm{d}}{\mathrm{d}x}(\mathrm{e}^{x})$

Here we have the simple exponent function e^x where constant $a = 1$.

Thus $\qquad\qquad \dfrac{\mathrm{d}}{\mathrm{d}x}(\mathrm{e}^{x}) = \mathrm{e}^{x}$

This verifies the important property of the exponent function, namely:

The exponent function e^x has a differential coefficient of e^x, which is therefore equal to itself.

We may now summarise differential coefficients of the more common functions:

y	$\dfrac{dy}{dx}$
$a x^n$	$a n x^{n-1}$
$\sin(\omega t + \alpha)$	$\omega \cos(\omega t + \alpha)$
$\cos(\omega t + \alpha)$	$-\omega \sin(\omega t + \alpha)$
$\log_e x$	$\dfrac{1}{x}$
e^{ax}	$a\,e^{ax}$

At this stage we may use the results in the table to differentiate many functions by 'recognition' rather than using the method of substitution.

For example, $$\frac{d}{d\theta}(\sin 3\theta) \;=\; 3\cos 3\theta$$

or $$\frac{d}{dx}(e^{-3x}) \;=\; -3e^{-3x}$$

However, if the functions to be differentiated are not similar to those shown in the table, perhaps because they are more complicated, the method of substitution should be used.

Example 13.8 which follows shows how both the methods of substitution and recognition may be used together to differentiate a more complicated function.

EXAMPLE 13.8

Find $\dfrac{dy}{dx}$ if $y = \sin^3 5x$

We have $\qquad\qquad y \;=\; (\sin 5x)^3$

Then $\qquad\qquad y \;=\; u^3 \qquad\qquad$ where $\quad u = \sin 5x$

$\therefore \qquad\qquad \dfrac{dy}{du} \;=\; 3u^2 \qquad\qquad$ and $\quad \dfrac{du}{dx} = 5\cos 5x$

But $\qquad \dfrac{dy}{dx} = \dfrac{dy}{du} \times \dfrac{du}{dx}$

$\therefore \qquad \dfrac{dy}{dx} = 3u^2 \times 5\cos 5x = 3\sin^2 5x \times 5\cos 5x$

$\therefore \qquad \dfrac{dy}{dx} = 15\sin^2 5x \cos 5x$

Exercise 13.2

Differentiate with respect to x:

1) $(3x + 1)^2$

2) $(2 - 5x)^3$

3) $(1 - 4x)^{1/2}$

4) $(2 - 5x)^{3/2}$

5) $\dfrac{1}{4x^2 + 3}$

6) $\sin(3x + 4)$

7) $\cos(2 - 5x)$

8) $\sin^2 4x$

9) $\dfrac{1}{\cos^3 7x}$

10) $\sin\left(2x + \dfrac{\pi}{2}\right)$

11) $\cos^3 x$

12) $\dfrac{1}{\sin x}$

13) $\log_e 9x$

14) $9\log_e\left(\dfrac{5}{x}\right)$

15) $\frac{1}{4}\log_e(2x - 7)$

16) $\dfrac{1}{e^x}$

17) $2e^{3x + 4}$

18) $\dfrac{1}{e^{2 - 8x}}$

19) Find $\dfrac{d}{dt}\left(\dfrac{1}{\sqrt[3]{1 - 2t}}\right)$

20) Find $\dfrac{d}{d\theta}[\sin(\frac{3}{4}\theta - \pi)]$

21) Find $\dfrac{d}{d\phi}\left(\dfrac{1}{\cos(\pi - \phi)}\right)$

22) Find $\dfrac{d}{dx}\left(\log_e \dfrac{1}{\sqrt{x}}\right)$

23) Find $\dfrac{d}{dt}(Be^{kt - b})$

24) Find $\dfrac{d}{dx}(\sqrt[3]{e^{1 - x}})$

DIFFERENTIATION OF A PRODUCT

If $y = u \times v$, where u and v are functions of x, we must use the formula

$$\frac{dy}{dx} = v\frac{du}{dx} + u\frac{dv}{dx}$$

EXAMPLE 13.9

Find $\dfrac{d}{dx}(x^3 \sin 2x)$

Let $y = x^3 \sin 2x$

Then $y = u \times v$ where $u = x^3$ and $v = \sin 2x$

$\therefore$ $\dfrac{du}{dx} = 3x^2$ and $\dfrac{dv}{dx} = 2\cos 2x$

But $\dfrac{dy}{dx} = v\dfrac{du}{dx} + u\dfrac{dv}{dx}$

$\therefore$ $\dfrac{dy}{dx} = (\sin 2x)3x^2 + x^3(2\cos 2x) = x^2(3\sin 2x + 2x\cos 2x)$

Then $\dfrac{d}{dx}(x^3 \sin 2x) = x^2(3\sin 2x + 2x\cos 2x)$

EXAMPLE 13.10

Differentiate $(x^2 + 1)\log_e x$ with respect to x

If we let $y = (x^2 + 1)\log_e x$ then the problem is to find $\dfrac{dy}{dx}$.

Then $y = u \times v$ where $u = x^2 + 1$ and $v = \log_e x$

$\therefore$ $\dfrac{du}{dx} = 2x$ and $\dfrac{dv}{dx} = \dfrac{1}{x}$

But $\dfrac{dy}{dx} = v\dfrac{du}{dx} + u\dfrac{dv}{dx}$

$\therefore$ $\dfrac{dy}{dx} = (\log_e x)2x + (x^2 + 1)\dfrac{1}{x} = 2x(\log_e x) + x + \dfrac{1}{x}$

DIFFERENTIATION OF A QUOTIENT

If $y = \dfrac{u}{v}$, where u and v are functions of x, we must use the formula

$$\frac{dy}{dx} = \frac{v\dfrac{du}{dx} - u\dfrac{dv}{dx}}{v^2}$$

EXAMPLE 13.11

Find $\dfrac{dy}{dx}$ if $y = \dfrac{e^{2x}}{x+3}$

We have $\qquad y = \dfrac{e^{2x}}{x+3}$

Let $\qquad y = \dfrac{u}{v} \qquad$ where $\quad u = e^{2x} \quad$ and $\quad v = x+3$

$$\therefore \qquad \frac{du}{dx} = 2e^{2x} \text{ and } \frac{dv}{dx} = 1$$

But $\qquad \dfrac{dy}{dx} = \dfrac{v\dfrac{du}{dx} - u\dfrac{dv}{dx}}{v^2}$

$\therefore \qquad \dfrac{dy}{dx} = \dfrac{(x+3)2e^{2x} - e^{2x} \times 1}{(x+3)^2}$

$\qquad\qquad = \dfrac{e^{2x}(2x + 6 - 1)}{(x+3)^2}$

$\qquad\qquad = \dfrac{(2x+5)e^{2x}}{(x+3)^2}$

EXAMPLE 13.12

Find $\dfrac{d}{d\theta}(\tan\theta)$

Let $\qquad y = \tan\theta$

or $\qquad y = \dfrac{\sin\theta}{\cos\theta}$

Then $\qquad y = \dfrac{u}{v} \qquad$ where $\quad u = \sin\theta \quad$ and $\quad v = \cos\theta$

$$\therefore \qquad \frac{du}{d\theta} = \cos\theta \quad\text{and}\quad \frac{dv}{d\theta} = -\sin\theta$$

But $\qquad \dfrac{dy}{d\theta} = \dfrac{v\,\dfrac{du}{d\theta} - u\,\dfrac{dv}{d\theta}}{v^2}$

$\therefore \qquad \dfrac{dy}{d\theta} = \dfrac{(\cos\theta)(\cos\theta) - (\sin\theta)(-\sin\theta)}{\cos^2\theta}$

$$= \frac{\cos^2\theta + \sin^2\theta}{\cos^2\theta}$$

Using the identity $\sin^2\theta + \cos^2\theta = 1$,

then $\qquad \dfrac{dy}{d\theta} = \dfrac{1}{\cos^2\theta} = \sec^2\theta$

Hence $\qquad \boxed{\dfrac{d}{d\theta}(\tan\theta) = \sec^2\theta} \quad$ provided that θ is in radians.

Exercise 13.3

1) Differentiate with respect to x:

  a) $x\sin x$ $\qquad\qquad$ **b)** $e^x \tan x$ $\qquad\qquad\qquad$ **c)** $x\log_e x$

2) Find $\dfrac{d}{dt}(\sin t \cos t)$ $\qquad\qquad$ **3)** Find $\dfrac{d}{d\theta}(\sin 2\theta \tan\theta)$

4) Find $\dfrac{d}{dm}(e^{4m}\cos 3m)$ $\qquad\qquad$ **5)** Find $\dfrac{d}{dx}(3x^2\log_e x)$

6) Find $\dfrac{d}{dt}[6e^{3t}(t^2 - 1)]$ $\qquad\qquad$ **7)** Find $\dfrac{d}{dz}[(z - 3z^2)\log_e z]$

8) Differentiate with respect to x:

  a) $\dfrac{x}{1 - x}$ $\qquad\qquad$ **b)** $\dfrac{\log_e x}{x^2}$ $\qquad\qquad$ **c)** $\dfrac{e^x}{\sin 2x}$

9) Find $\dfrac{d}{dz}\left(\dfrac{z + 2}{3 - 4z}\right)$ $\qquad\qquad$ **10)** Find $\dfrac{d}{dt}\left(\dfrac{\cos 2t}{e^{2t}}\right)$

11 Find $\dfrac{d}{d\theta}(\cot\theta)$. $\quad\left(Hint:\text{ Use the identity } \cot\theta = \dfrac{\cos\theta}{\sin\theta}\right)$

14

Practical Problems Solved Using Differentiation

Distance, time, velocity and acceleration – linear and angular – exponential graphs – growth and decay problems – turning points – maximum and minimum problems.

INTRODUCTION

At last some practical applications of differentiation. Linear and rotary movement is met constantly in technology. Also growth and decay problems – electrical circuits, population and radioactivity, and even compound interest in money. Finally an extension to our knowledge of graphs – turning points and how they may be used in solving problems involving maxima and minima.

DISTANCE, TIME, VELOCITY AND ACCELERATION

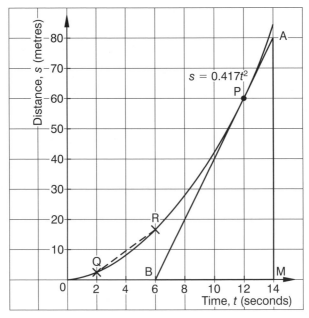

Fig. 14.1

Suppose that a vehicle starts from rest and travels 60 metres in 12 seconds. The average velocity may be found by dividing the total distance travelled by the total time taken, that is $\frac{60}{12} = 5$ m/s. This is **not** the *instantaneous* velocity at any instant but is the *average velocity* over the distance travelled in 12 seconds.

Figure 14.1 shows a graph of distance s against time t. The average velocity over a period is given by the gradient of the chord which meets the curve at the extremes of the period. Thus in the diagram the gradient of the dotted chord QR gives the average velocity between $t = 2$ s and $t = 6$ s. It is found to be $\frac{13}{4} = 3.25$ m/s.

The velocity, at any point, is the rate of change of s with respect to t and may be found by finding the gradient of the curve at that point. In other words:

> The rate of change of distance with respect to time is called **velocity** and is given by the **gradient of the distance–time graph** at any point.

In mathematical notation that is given by $\dfrac{ds}{dt}$. Thus: velocity $v = \dfrac{ds}{dt}$.

Suppose we know that the relationship between s and t is

$$s = 0.417\, t^2$$

Then velocity $\qquad\qquad\qquad v = \dfrac{ds}{dt} = 0.834\, t$

thus when $t = 12$ seconds, $\qquad v = 0.834 \times 12 = 10$ m/s.

This result may be found graphically, Fig. 14.1, by drawing the tangent to the curve of s against t at point P and constructing a suitable right-angled $\triangle$ABM.

$\therefore$ velocity at P $= \dfrac{AM}{BM} = \dfrac{80}{8} = 10$ m/s verifying the theoretical result.

Similarly, rate of change of velocity with respect to time is called **acceleration** and is given by the **gradient of the velocity–time graph** at any point. In mathematical notation $\dfrac{dv}{dt}$. Thus acceleration $a = \dfrac{dv}{dt}$.

The above reasoning was applied to linear motion, but it could also have been used for angular motion. The essential difference is that distance s is replaced by angle turned through θ rad.

Both sets of results are summarised in Fig. 14.2.

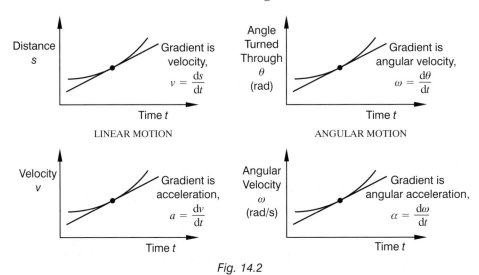

Fig. 14.2

EXAMPLE 14.1

The research and development department of a company is developing a test rig to verify a theory for advanced dimensional analysis. A remotely controlled model has been programmed to move according to the equation $s = 2t^3 - 9t^2 + 12t + 6$ where s metres is the distance moved by the model at a time t seconds. We have been asked to find:

a) its velocity after 3 s
b) its acceleration after 3 s
c) when the velocity is zero.

We have the   displacement $s = 2t^3 - 9t^2 + 12t + 6$

$\therefore$ the   velocity $v = \dfrac{ds}{dt} = 6t^2 - 18t + 12$

and the   acceleration $a = \dfrac{dv}{dt} = 12t - 18$

a) When $t = 3$ the velocity $v = \dfrac{ds}{dt} = 6(3)^2 - 18(3) + 12 = 12 \, \text{m/s}$

b) When $t = 3$ the acceleration $a = \dfrac{dv}{dt} = 12(3) - 18 = 18 \, \text{m/s}^2$

c) When velocity is zero, that is when $v = \dfrac{ds}{dt} = 0$

Then $\qquad\qquad 6t^2 - 18t + 12 = 0$

$\therefore\qquad\qquad\qquad\quad t^2 - 3t + 2 = 0$ by dividing through by 6

$\therefore\qquad\qquad\qquad (t-1)(t-2) = 0$ by factorising

$\therefore\qquad$ either $\qquad\quad t - 1 = 0$ or $t - 2 = 0$

$\therefore\qquad$ either $\qquad\qquad t = 1\,\mathrm{s}$ or $t = 2\,\mathrm{s}$

EXAMPLE 14.2

A displacement transducer has been fitted to a disc attached to a spindle, in order to measure the angle through which the spindle turns. After a period of observation of the machine tool being investigated, it has been established that the angular motion is governed by the relationship $\theta = 20 + 5t^2 - t^3$ where θ radians is the angle turned through at a time t seconds. We have been asked to find

a) the angular velocity when $t = 2$ seconds
b) the value of t when angular deceleration is $4\,\mathrm{rad/s^2}$.

Now the angular displacement $\qquad\qquad\qquad \theta = 20 + 5t^2 - t^3$

$\therefore$ the angular velocity $\qquad\qquad \omega = \dfrac{d\theta}{dt} = 10t - 3t^2$

and the angular acceleration $\qquad \alpha = \dfrac{d\omega}{dt} = 10 - 6t$

a) When $t = 2$ then the angular velocity

$$\omega = \frac{d\theta}{dt} = 10(2) - 3(2)^2 = 8\,\mathrm{rad/s}$$

b) An angular deceleration of $4\,\mathrm{rad/s^2}$ may be called an angular acceleration of $-4\,\mathrm{rad/s^2}$

$\therefore$ when $\alpha = \dfrac{d\omega}{dt} = -4$ then $-4 = 10 - 6t$

$\qquad\qquad\qquad\qquad$ or $\qquad t = 2.33$ seconds

Exercise 14.1

1) If $s = 10 + 50t - 2t^2$, where s metres is the distance travelled in t seconds by a body, what is the velocity of the body after 2 seconds?

2) If $v = 5 + 24t - 3t^2$ where v m/s is the velocity of a body at a time t seconds, what is the acceleration when $t = 3$?

3) A body moves s metres in t seconds where $s = t^3 - 3t^2 - 3t + 8$. Find:

   a) its velocity at the end of 3 seconds,

   b) when its velocity is zero,

   c) its acceleration at the end of 2 seconds,

   d) when its acceleration is zero.

4) A body moves s metres in t seconds, where $s = \dfrac{1}{t^2}$. Find the velocity and acceleration after 3 seconds.

5) The distance s metres travelled by a falling body starting from rest after a time t seconds is given by $s = 5t^2$. Find its velocity after 1 second and after 3 seconds.

6) The distance s metres moved by the end of a lever after a time t seconds is given by the formula $s = 6t^2$. Find the velocity of the end of the lever when it has moved a distance $\frac{1}{2}$ metre.

7) The angular displacement θ radians of the spoke of a wheel is given by the expression $\theta = \frac{1}{2}t^4 - t^3$ where t seconds is the time. Find:

   a) the angular velocity after 2 seconds,

   b) the angular acceleration after 3 seconds,

   c) when the angular acceleration is zero.

8) An angular displacement θ radians in time t seconds is given by the equation $\theta = \sin 3t$. Find:

   a) the angular velocity when $t = 1$ second,

   b) the smallest positive value of t for which the angular velocity is 2 rad/s,

   c) the angular acceleration when $t = 0.5$ seconds,

   d) the smallest positive value of t for which the angular acceleration is 9 rad/s^2.

9) A mass of 5000 kg moves along a straight line so that the distance s metres travelled in a time t seconds is given by $s = 3t^2 + 2t + 3$. If v m/s is its velocity and m kg is its mass, then its kinetic energy is given by the formula $\frac{1}{2}mv^2$. Find its kinetic energy at a time $t = 0.5$ seconds, remembering that the joule (J) is the unit of energy.

EXPONENT CURVES – GROWTH AND DECAY

Using a Time Base – the Curve of e^t

In technology reference to the exponent curves is usually based on time. Thus the curves are plotted on a time t base and not an x base.

Practical applications based on growth and decay start from a given instant. This means we are dealing with actual (or real) times which, in numerical terms, are positive values of t. Thus only portions of the curves to the right of the vertical axis are generally used, as shown in Fig. 14.3.

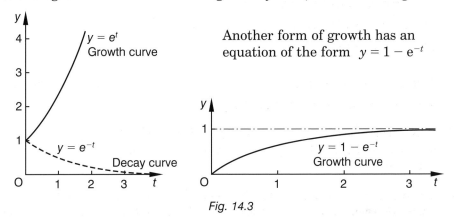

Another form of growth has an equation of the form $y = 1 - e^{-t}$

Fig. 14.3

General Expressions of Growth and Decay

When used to solve problems the equations are modified to the following general forms

$$y = ae^{bt}, \quad y = ae^{-bt}, \quad \text{and} \quad y = a(1 - e^{-bt})$$

where a and b are constants which affect the initial value of y and also the rate of growth or decay. Sketches of the curves and some typical applications of each are shown in Fig. 14.4.

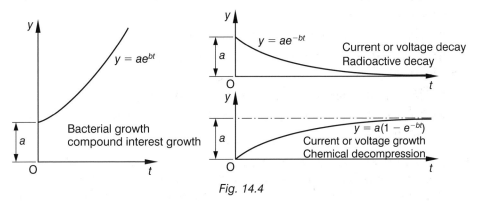

Fig. 14.4

EXAMPLE 14.3

The instantaneous e.m.f. in an inductive circuit is given by the expression $100.e^{-4t}$ volts, where t is time in seconds. Plot the graph of the e.m.f. for values of t from 0 to 0.5 seconds, and use the graph to find:

a) the value of the e.m.f. when $t = 0.25$ seconds, and
b) the rate of change of the e.m.f. when $t = 0.1$ seconds.

Check this result using the method of differentiation.

To plot a graph we need the e.m.f. for values of t,
say $t = 0$, 0.1, 0.2, 0.3, 0.4, 0.5
A suitable calculator sequence for e.g. $t = 0.2$ would be

$\boxed{\text{AC}}\ \boxed{100}\ \boxed{e^x}\ \boxed{(}\ \boxed{-4}\ \boxed{\times}\ \boxed{0.2}\ \boxed{)}\ \boxed{=}$ giving 45 to two s.f.

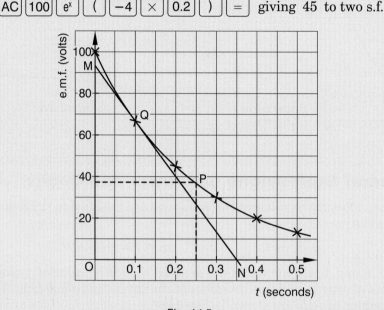

Fig. 14.5

a) The point P on the curve is at 0.25 seconds shown on the t scale and the corresponding value of e.m.f. can be read directly from the vertical axis scale. The value is 37 volts.

b) The point Q on the graph is at 0.1 seconds. Now the rate of change of the curve at Q is given by the gradient of the tangent at Q. This gradient may be found by constructing a suitable right angled triangle such as MNO in Fig. 14.5, and finding the ratio $\dfrac{MO}{ON}$.

Hence the gradient at Q

$$= \frac{MO}{ON} = \frac{92 \text{ volts}}{0.34 \text{ seconds}} = 270 \text{ volts per second}$$

According to the sign convention a line sloping downwards from left to right has a negative gradient.

Hence the gradient at Q is -270 volts per second, which means that the rate of change of the curve at Q is -270 volts per second.

This is the same as saying that the e.m.f. at $t = 0.1$ seconds is decreasing at the rate of 270 volts per second.

A Check Using Differentiation

The rate of change of the e.m.f. at any instant is given by the gradient of the e.m.f. – time curve and using the calculus this is denoted by $\dfrac{d}{dt}$ (e.m.f.).

Since we have

$$\text{e.m.f.} = 100.e^{-4t}$$

then

$$\frac{d}{dt}(\text{e.m.f.}) = 100(-4)e^{-4t}$$

$$= -400e^{-4t}$$

Hence when $t = 0.1$ seconds

then the required rate of change $= -400.e^{-4(0.1)}$

$$= -400.e^{-0.4}$$

$$= -268 \text{ volts/second}$$

This verifies the approximate value of -270 volts/second obtained by use of the graph.

EXAMPLE 14.4

The formula $i = 2(1 - e^{-10t})$ gives the relationship between the instantaneous current i amperes and the time t seconds in an inductive circuit. Plot a graph of i against t, taking values of t from 0 to 0.3 seconds at intervals of 0.05 seconds. Hence find:

a) the initial rate of growth of the current i when $t = 0$, and
b) the time taken for the current to increase from 1 to 1.6 amperes.

Verify these results using theoretical methods.

To plot a graph we need i for values of t,
say $t = 0$, 0.05, 0.10, 0.15, 0.20, 0.25, 0.30
A suitable calculator sequence for e.g. $t = 0.20$ using two sets of brackets
would be

giving 1.73 to 3 s.f. The curve is shown plotted in Fig. 14.6.

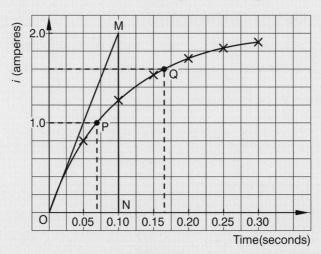

Fig. 14.6

a) When $t = 0$ the initial rate of growth will be given by the gradient of
the tangent at O. The tangent at O is the line OM and its gradient
may be found by using a suitable right angled triangle MNO and
finding the ratio $\dfrac{\text{MN}}{\text{ON}}$.

Hence the initial rate of growth of $i = \dfrac{\text{MN}}{\text{ON}} = \dfrac{2 \text{ amperes}}{0.1 \text{ seconds}}$

$= 20$ amperes per second

b) The point P on the curve corresponds to a current of 1.0 amperes
and the time at which this occurs may be read from the t scale and is
0.07 seconds.

Similarly point Q corresponds to a current of 1.6 ampere and occurs
at 0.16 seconds.

Hence the time between P and Q is 0.16–0.07 = 0.09 seconds.

This means that the time for the current
to increase from 1 to 1.6 amperes is 0.09 seconds.

Solution Using Theoretical Methods

a) The rate of growth of current is another way of stating a positive rate of change of current. At any instant this is given by the gradient of the current–time curve and using the calculus this is denoted by $\dfrac{d}{dt}$(current).

Since we have   current $= 2(1 - e^{-10t})$

or $i = 2 - 2.e^{-10t}$

then $\dfrac{di}{dt} = -2(-10)e^{-10t} = 20e^{-10t}$

Hence when $t = 0$ seconds

the rate of change of current $= 20.e^{-10(0)}$

$= 20.e^0$

$= 20$ A/s   (since $e^0 = 1$)

This verifies the value obtained by use of the graph.

b) We have $i = 2(1 - e^{-10t})$

hence when $i = 1.6$ A then

$1.6 = 2(1 - e^{-10t})$

and rearranging $e^{-10t} = 1 - \dfrac{1.6}{2} = 0.2$

and rewriting this equation in logarithmic form

$-10t = \log_e 0.2$

$\therefore$ $t = -\dfrac{1}{10}\log_e 0.2 = 0.161$ s

Also when $i = 1.0$ A then

$1 = 2(1 - e^{-10t})$

and rearranging $e^{-10t} = 1 - \dfrac{1}{2} = 0.5$

and rewriting this equation in logarithmic form

$-10t = \log_e 0.5$

$\therefore$ $t = -\dfrac{1}{10}\log_e 0.5 = 0.069$ s

required time $=$ (time to increase to 1.6 A) $-$ (time to increase to 0.1 A)

$= 0.161 - 0.069 = 0.092$ s

This verifies the more approximate value obtained by use of the graph.

Exercise 14.2

Check the results obtained graphically by using calculus.

1) For a constant pressure process on a certain gas the formula connecting the absolute temperature T and the specific entropy s is $T = 24.e^{3s}$. Plot a graph of T against s taking values of s equal to 1.000, 1.033, 1.066, 1.100, 1.133, 1.166 and 1.200. Use the graph to find the value of:

   a) T when $s = 1.09$

   b) s when $T = 700$

2) The equation $i = 2.4e^{-6t}$ gives the relationship between the instantaneous current, i mA, and the time, t seconds. Plot a graph of i against t for values of t from 0 to 0.6 seconds at 0.1 second intervals. Use the curve obtained to find the rate at which the current is decreasing when $t = 0.2$ seconds.

3) In a capacitive circuit the voltage v and the time t seconds are connected by the relationship $v = 240(1 - e^{-5t})$. Draw the curve of v against t for values of $t = 0$ to $t = 0.7$ seconds at 0.1 second intervals.

   Hence find:

   a) the time when the voltage is 140 volts,

   b) the initial rate of growth of the voltage when $t = 0$.

4) A radioactive material decays according to the law $N = N_i(e^{-0.0693t})$ where N is the amount of radioactivity remaining, N_i is the initial amount of radioactivity, and t years is the time taken to decay from N_i to N. Plot a graph of N (this will be in terms of N_i e.g. $0.4N_i$) against t for values of $t = 0$ to $t = 30$ years at 5 yearly intervals. Hence find the half life of the material (that is when half the radioactivity has decayed or $N = \frac{1}{2}N_i$).

TURNING POINTS

When we look at the features of a mathematical curve we appreciate first the overall shape and layout. Other important features are where it cuts the reference axes and the turning points – these are especially of interest when using calculus, as you will see.

Consider the graph of $y = x^3 + 3x^2 - 9x + 6$.

To plot the graph we need values of y for values of x say from -4 to $+3$. The simplest way is to nest the equation giving $x + 3)x - 9)x + 6$ and then use the calculator remembering to work from left to right. So for $x = -4$ we have

$$\boxed{AC}\;\boxed{-4}\;\boxed{Min}\;\boxed{+}\;\boxed{3}\;\boxed{=}\;\boxed{\times}\;\boxed{RCL}\;\boxed{M}\;\boxed{-}\;\boxed{9}\;\boxed{=}$$
$$\boxed{\times}\;\boxed{RCL}\;\boxed{M}\;\boxed{+}\;\boxed{6}\;\boxed{=}\;\text{giving } 26$$

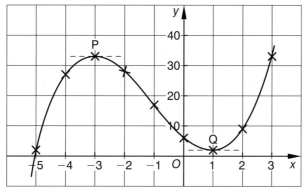

Fig. 14.7

The points P and Q are called turning points since the gradient of the tangent is zero at each of them. Also P is called a maximum and Q is called a minimum. It will be seen from the graph that the value of y at P is not the greatest value that y can have – nor the value of y at Q the least.

The terms maximum and minimum values apply only in the vicinity of the turning points and not to the values of y in general.

So at P when $x = -3$, we have a maximum value of y of 33 and at Q when $x = 1$, we have a minimum value of y of 1.

MAXIMA AND MINIMA

An important use of calculus is its application to finding maxima and minima, especially to a wide variety of engineering problems.

It is not always convenient to draw the full graph to find the turning points. At the turning points the gradient (slope) is zero, i.e. $\dfrac{dy}{dx} = 0$ in calculus notation (Fig. 14.8). As you will see this enables us to find the turning points.

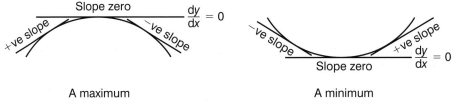

Fig. 14.8

It is usually necessary to determine if a turning point is a maximum or minimum. Figure 14.8 shows how the gradient of a curve changes in the vicinity of a turning point.

EXAMPLE 14.5

Check the results for maximum and minimum values of y, found previously be a graphical means, given that $y = x^3 + 3x^2 - 9x + 6$.

We have
$$y = x^3 + 3x^2 - 9x + 6$$

∴
$$\frac{dy}{dx} = 3x^2 + 6x - 9$$

At a turning point
$$\frac{dy}{dx} = 0$$

∴
$$3x^2 + 6x - 9 = 0$$

We will now simplify by dividing through by 3, and solve the quadratic equation using factors:

thus
$$x^2 + 2x - 3 = 0$$

or
$$(x - 1)(x + 3) = 0$$

∴ either
$$(x - 1) = 0 \quad \text{or} \quad x + 3 = 0$$

so either
$$x = 1 \quad \text{or} \quad x = -3$$

and these confirm the turning point positions shown in Fig. 14.7.

Test for Maximum or Minimum

At the turning point where $x = 1$, we know that $\dfrac{dy}{dx} = 0$, i.e. zero slope,

and using a value of x slightly less than 1, say $x = 0.5$,

gives $\dfrac{dy}{dx} = 3(0.5)^2 + 6(0.5) - 9 = -5.25$, i.e. negative slope,

and using a value of x slightly greater than 1, say $x = 1.5$,

gives $\dfrac{dy}{dx} = 3(1.5)^2 + 6(1.5) - 9 = +6.75$, i.e. positive slope.

These results are best shown by means of a diagram (Fig. 14.9) which indicates clearly that when $x = 1$ we have a minimum.

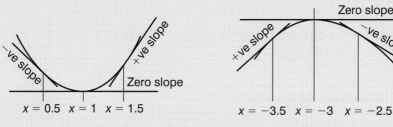

Fig. 14.9

The minimum value of y may be found by substituting $x = 1$ into the given equation. Hence

$$y_{\min} = (1)^3 + 3(1)^2 - 9(1) + 6 = 1$$

Similarly for the turning point at $x = -3$, at values of $x = -3.5$ (slightly below -3) and $x = -2.5$ (slightly above -3) we obtain the result shown in Fig. 14.9 which indicates that when $x = -3$ we have a maximum.

The maximum value of y may be found by substituting $x = -3$ into the given equation. Hence

$$y_{\max} = (-3)^3 + 3(-3)^2 - 9(-3) + 6) = 33$$

APPLICATIONS IN TECHNOLOGY

In order to find maxima and minima using calculus we need an equation connecting the quantity for which a maximum or minimum is required in terms of another variable. It may well be necessary to form this equation and it may help to draw a diagram representing the problem.

EXAMPLE 14.6

An electric current is represented by $i = 5 \sin \theta$ where θ is in radians. Find the maximum value of i between $\theta = 0$ and π radians.

We have
$$i = 5 \sin \theta$$

then
$$\frac{di}{d\theta} = 5 \cos \theta$$

At a turning point
$$\frac{di}{d\theta} = 0$$

i.e.
$$5 \cos \theta = 0$$

or
$$\cos \theta = 0$$

giving
$$\theta = \frac{\pi}{2} \text{ rad}$$

This is the only solution between 0 and π radians.

Test for Maximum or Minimum

At the turning point where $\theta = \dfrac{\pi}{2} = 1.57 \text{ rad}$ we know that

$\dfrac{di}{d\theta} = 0$, i.e. zero slope and using a value of θ slightly less than 1.57,

say $\theta = 1.50$, gives $\dfrac{di}{d\theta} = 5 \cos 1.50 = 0.35$, i.e. positive slope

and using a value of θ slightly greater than 1.57, say $\theta = 1.60$, gives

$\dfrac{di}{d\theta} = 5 \cos 1.60 = -0.15$, i.e. negative slope.

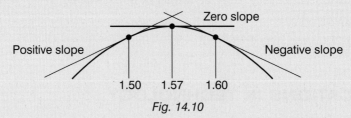

Fig. 14.10

Figure 14.10 indicates a maximum at $\theta = \dfrac{\pi}{2} \text{ rad}$, since $\sin \dfrac{\pi}{2} = 1$

then the maximum value of current $i_{\max} = 5 \sin \dfrac{\pi}{2} = 5$

EXAMPLE 14.7

A rectangular sheet of metal 360 mm by 240 mm has four equal squares cut out at the corners. The sides are then turned up to form a rectangular box. Find the length of the sides of the squares cut out so that the volume of the box may be as great as possible and find this maximum volume.

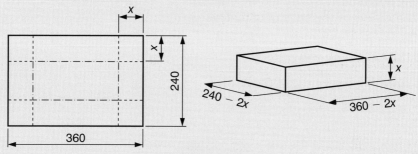

Fig. 14.11

Let the length of the side of each cut away square be x mm as shown in Fig. 14.11.

Hence the volume is

$$V = x(240 - 2x)(360 - 2x)$$
$$= 4x^3 - 1200x^2 + 86\,400x$$

$\therefore$

$$\frac{dV}{dx} = 12x^2 - 2400x + 86\,400$$

At a turning point

$$\frac{dV}{dx} = 0$$

$\therefore \qquad 12x^2 - 2400x + 86\,400 = 0$

or $\qquad x^2 - 200x + 7200 = 0 \quad$ by dividing through by 12

Now this is a quadratic equation which does not factorise easily so we will have to solve using the formula for the standard quadratic

$ax^2 + bx + c = 0 \quad$ giving $\quad x = \dfrac{-b \pm \sqrt{b^2 - 4ac}}{2a}$

Hence the solution of our equation is

$$x = \frac{-(-200) \pm \sqrt{(-200)^2 - 4 \times 1 \times 7200}}{2 \times 1}$$

$\therefore$ either $\qquad x = 152.9 \quad$ or $\quad x = 47.1$

However, from the physical sizes of the sheet, it is not possible for x to be 152.9 mm (since one side is only 240 mm long) so we reject this solution. Hence $x = 47.1$ mm.

We will leave you to check that the turning point at $x = 47.1$ is a maximum.

It only remains to find the maximum volume by substituting $x = 47.1$ into the equation for V. Therefore

$$V_{max} = 47.1(240 - 2 \times 47.1)(360 - 2 \times 47.1)$$
$$= 1.825 \times 10^6 \, \text{mm}^3$$

EXAMPLE 14.8

A cylinder with an open top has a capacity of $2\,\mathrm{m}^3$ and is made from sheet metal. Neglecting any overlaps at the joints find the dimensions of the cylinder so that the amount of sheet steel used is a minimum.

Let the height of the cylinder be h metres and the radius of the base be r metres as shown in Fig. 14.12.

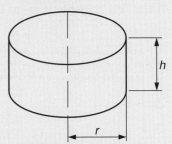

Now the total area of metal

    = area of base + area of curved side

$$A = \pi r^2 + 2\pi r h$$

Fig. 14.12

We cannot proceed to differentiate as there are two variables on the right-hand side of the equation. It is possible, however, to find a connection between r and h using the fact that the volume is $2\,\mathrm{m}^3$.

Now   Volume of a cylinder $= \pi r^2 h$

$\therefore$ $2 = \pi r^2 h$

from which $h = \dfrac{2}{\pi r^2}$

We may now substitute for h in the equation for A.

$\therefore$ $A = \pi r^2 + 2\pi r \left(\dfrac{2}{\pi r^2} \right)$

$$= \pi r^2 + \dfrac{4}{r}$$

$$= \pi r^2 + 4r^{-1}$$

$\therefore$ $\dfrac{\mathrm{d}A}{\mathrm{d}r} = 2\pi r - 4r^{-2}$

Now for a turning point $\dfrac{\mathrm{d}A}{\mathrm{d}r} = 0$

or $2\pi r - 4r^{-2} = 0$

$\therefore$ $2\pi r - \dfrac{4}{r^2} = 0$

$\therefore$ $2\pi r = \dfrac{4}{r^2}$

$\therefore$ $r^3 = \dfrac{2}{\pi} = 0.637$

$$r = \sqrt[3]{0.637} = 0.860$$

Again we leave you to check that the turning point at $r = 0.86$ is a minimum.

Hence $r = 0.86\,\text{m}$ makes A a minimum as required.

We may find the corresponding value of h by substituting $r = 0.86$ into the equation found previously for h in terms of r

$$h = \frac{2}{\pi(0.86)^2} = 0.86$$

Hence for the minimum amount of metal to be used the radius is $0.86\,\text{m}$ and the height is also $0.86\,\text{m}$.

Exercise 14.3

1) Find the maximum and minimum values of:

   a) $y = 2x^3 - 3x^2 - 12x + 4$ **b)** $y = x^3 - 3x^2 + 4$

   c) $y = 6x^2 + x^3$

2) Given that $y = 60x + 3x^2 - 4x^3$, calculate:

   a) the gradient of the tangent to the curve of y at the point where $x = 1$,

   b) the value of x for which y has its maximum value,

   c) the value of x for which y has its minimum value.

3) Calculate the coordinates of the points on the curve

$$y = x^3 - 3x^2 - 9x + 12$$

at each of which the tangent to the curve is parallel to the x-axis.

4) A curve has the equation $y = 8 + 2x - x^2$. Find:

   a) the value of x for which the gradient of the curve is 6,

   b) the value of x which gives the maximum value of y,

   c) the maximum value of y.

5) The curve $y = 2x^2 + \dfrac{k}{x}$ has a gradient of 5 when $x = 2$.

   Calculate:

   a) the value of k,   **b)** the minimum value of y.

6) Find the maximum and minimum values of the voltage v given by the expression $v = 3(\cos\theta) + 7$ between, and including, values of $\theta = 0$ and $\theta = 2\pi$ radians.

7) In an electric circuit the number of heat units, H, produced when a current, i, is flowing is given by $H = Ei - Ri^2$ where E is the e.m.f. and R is the resistance. Find the maximum heat that can be produced when $E = 8$ and $R = 2$.

8) The power, p watts, available from the rotor of an undershot waterwheel is given by $p = \dot{m}(v_j - u)u$. The water mass flow rate $\dot{m} = 7.85$ kg/s and the velocity of the vanes $v_j = 25$ m/s. Find the velocity u of the vanes for maximum power. Find also the value of the maximum power.

9) From a rectangular sheet of metal measuring 120 mm by 75 mm equal squares of side x are cut from each of the corners. The remaining flaps are then folded upwards to form an open box.

Show that the volume of the box is given by

$$V = 9000x - 390x^2 + 4x^3$$

Find the value of x such that the volume is a maximum.

10) An open rectangular tank of height h metres with a square base of side x metres is to be constructed, so that it has a capacity of 500 cubic metres. Show that the surface area of the four walls and the base will be $\dfrac{2000}{x} + x^2$ square metres. Find the value of x for this expression to be a minimum.

11) The volume of a cone is given by the formula $V = \frac{1}{3}\pi r^2 h$, where h is the height of the cone and r its radius. If $h = 6 - r$, calculate the value of r for which the volume is a maximum.

12) A box without a lid has a square base of side x mm and rectangular sides of height h mm. It is made from 10 800 mm² of sheet metal of negligible thickness. Show that $h = \dfrac{10\,800 - x^2}{4x}$ and that the volume of the box is $(2700x - \frac{1}{4}x^3)$. Hence calculate the maximum volume of the box.

13) A cylindrical lemonade can made from thin metal has to hold 0.5 litres. Find its dimensions if the area of metal used is a minimum.

14) A cooling tank is to be made with the trapezoidal section as shown:

Its cross-sectional area is to be 300 000 mm². Show that the width of material needed to form, from one sheet, the bottom and folded-up sides is $w = \dfrac{300\,000}{h} + 1.828h$. Hence find the height h of the tank so that the width of material needed is a minimum.

15) A cylindrical cup is to be drawn from a disc of metal of 50 mm diameter. Assuming that the surface area of the cup is the same as that of the disc find the dimensions of the cup so that its volume is a maximum.

Integration as a reverse of differentiation –
standard integrals x^n, $\sin ax$, $\cos ax$, $1/x$, e^{ax} – indefinite and definite
integrals – integration by substitution – change of limits – integration by parts.

INTRODUCTION

We will revise our ideas of integration as the reverse of differentiation.
This enables us to use standard integrals to perform indefinite integration
and evaluate definite integrals.

We shall also be introduced to integration by substitution and by parts.

Integration is our second 'tool' in calculus and it will be used in later
chapters to find areas, volumes, centroids, second moments of area,
moments of inertia and mean and root mean square (r.m.s) values.

INTEGRATION AS THE REVERSE OF DIFFERENTIATION

We have previously discovered how to obtain the differential coefficients of
various functions. Our objective in this section is to find out how to
reverse the process. That is, being given the differential coefficient of a
function we try to discover the original function.

if $$y = \frac{x^4}{4}$$

then $$\frac{dy}{dx} = x^3$$

or we may write $$dy = x^3 \, dx$$

The expression $x^3 \, dx$ is called the differential of $\dfrac{x^4}{4}$.

Reversing the process of differentiation is called *integration*.

It is indicated by using the integration sign $\int$ in front of the differential.

Thus, if: $$dy = x^3 \, dx$$

then reversing the process $$y = \int x^3 \, dx = \frac{x^4}{4}$$

Now $\dfrac{x^4}{4}$ is called the integral of $x^3 \, dx$.

Similarly

$$\int x^n \, dx = \frac{x^{n+1}}{n+1}$$

Remember
'increase the index by one, and then divide by the new index'

This rule applies to all indices whether positive, negative or fractional except for $\int x^{-1} \, dx = \int \frac{1}{x} \, dx$ because since $\frac{d}{dx} (\log_e x) = \frac{1}{x}$ the reverse of this gives:

$$\int \frac{1}{x} \, dx = \log_e x$$

Since $\frac{d}{dt} \sin(\omega t + \alpha) = \omega \cos(\omega t + \alpha)$ the reverse of this gives:

$$\int \cos(\omega t + \alpha) \, dt = \frac{1}{\omega} \sin(\omega t + \alpha)$$

Similarly

$$\int \sin(\omega t + \alpha) \, dt = -\frac{1}{\omega} \cos(\omega t + \alpha)$$

Since $\frac{d}{dx} e^{ax} = a e^{ax}$ the reverse of this gives:

$$\int e^{ax} \, dx = \frac{1}{a} e^{ax}$$

Summarising:

$$\int x^n \, dx = \frac{x^{n+1}}{n+1}$$

$$\int \cos(\omega t + \alpha) \, dt = \frac{1}{\omega} \sin(\omega t + \alpha)$$

$$\int \sin(\omega t + \alpha) \, dt = -\frac{1}{\omega} \cos(\omega t + \alpha)$$

$$\int \frac{1}{x} \, dx = \log_e x$$

$$\int e^{ax} \, dx = \frac{1}{a} e^{ax}$$

INDEFINITE INTEGRALS

An **indefinite** integral is an integral *without limits* and the solution must therefore contain a constant of integration.

EXAMPLE 15.1

Find $\int (x^2 + 2x - 3)\,dx$

$$\int (x^2 + 2x - 3)\,dx = \frac{x^3}{3} + 2\frac{x^2}{2} - 3x + k$$

$$= \frac{x^3}{3} + x^2 - 3x + k$$

EXAMPLE 15.2

Find $\int (\sin 7\theta + 2 \cos 5\theta)\,d\theta$

$$\int (\sin 7\theta + 2 \cos 5\theta)\,d\theta = -\tfrac{1}{7}\cos 7\theta + \tfrac{2}{5}\sin 5\theta + k$$

EXAMPLE 15.3

Find $\int \left(e^{6t} - \dfrac{1}{e^{3t}} \right) dt$

$$\int \left(e^{6t} - \frac{1}{e^{3t}} \right) dt = \int (e^{6t} - e^{-3t})\,dt$$

$$= \tfrac{1}{6}e^{6t} - \frac{1}{(-3)}e^{-3t} + k$$

$$= \tfrac{1}{6}e^{6t} + \tfrac{1}{3}e^{-3t} + k$$

DEFINITE INTEGRALS

A **definite** integral *has limits*. Square brackets are used to indicate that we have completed the actual integration and will next be substituting the values of the limits.

EXAMPLE 15.4

Evaluate $\displaystyle\int_2^3 (1 + \cos 2\phi)\, d\phi$

We should remember that when limits are substituted into trigonometrical functions they represent radian values (not degrees).

$$\int_2^3 (1 + \cos 2\phi)\, d\phi = [\phi + \tfrac{1}{2}\sin 2\phi]_2^3$$
$$= \{3 + \tfrac{1}{2}\sin(2 \times 3)\} - \{2 + \tfrac{1}{2}\sin(2 \times 2)\}$$
$$= 3 + \tfrac{1}{2}\sin 6 - 2 - \tfrac{1}{2}\sin 4$$
$$= 1.24$$

EXAMPLE 15.5

Evaluate $\displaystyle\int_0^1 5(e^{2t} - e^{-2t})\, dt$

$$\int_0^1 5(e^{2t} - e^{-2t})\, dt = 5\int_0^1 (e^{2t} - e^{-2t})\, dt$$
$$= 5\left[\tfrac{1}{2}e^{2t} - \frac{1}{(-2)}e^{-2t}\right]_0^1$$
$$= \tfrac{5}{2}\{(e^{2 \times 1} + e^{-2 \times 1}) - (e^{2 \times 0} + e^{-2 \times 0})\}$$
$$= \tfrac{5}{2}(e^2 + e^{-2} - e^0 - e^0)$$
$$= \tfrac{5}{2}(7.39 + 0.14 - 1 - 1) = 13.8$$

EXAMPLE 15.6

Evaluate $\displaystyle\int_1^2 \left(\frac{1}{x}\right) dx$

$$\int_1^2 \left(\frac{1}{x}\right) dx = \left[\log_e x\right]_1^2$$
$$= (\log_e 2 - \log_e 1)$$
$$= \log_e \frac{2}{1} = \log_e 2 = 0.693$$

Exercise 15.1

Find:

1) $\int \left(5x^2 + 2x - \dfrac{4}{x^2} \right) dx$

2) $\int \left(\sqrt{x} + \dfrac{1}{\sqrt{x}} \right) dx$

3) $\int \sin \dfrac{x}{3} \, dx$

4) $\int 5 \cos 3\theta \, d\theta$

5) $\int (1 + \sin \tfrac{2}{3}\phi) \, d\phi$

6) $\int \left(\cos \dfrac{\theta}{2} - \sin \dfrac{3\theta}{2} \right) d\theta$

7) $\int (2t + \sin 2t) \, dt$

8) $\int e^{3x} \, dx$

9) $\int e^{-0.5u} \, du$

10) $\int (3e^{2t} - 2e^t) \, dt$

11) $\int (e^{-x/2} + e^{3x/2}) \, dx$

12) $\int \left(\sec^2 x + \dfrac{1}{x} \right) dx$

Evaluate:

13) $\displaystyle\int_1^2 (x^3 + 4) \, dx$

14) $\displaystyle\int_0^2 \left(\dfrac{x^2 + x^3}{x} \right) dx$

15) $\displaystyle\int_0^1 \dfrac{2 \cos x}{3} \, dx$

16) $\displaystyle\int_0^{\pi/2} 3 \sin 4\phi \, d\phi$

17) $\displaystyle\int_{\pi/6}^{\pi/3} 2 \sin \dfrac{2t}{3} \, dt$

18) $\displaystyle\int_{-\pi/2}^{\pi/2} \sin 2\theta \, d\theta$

19) $\displaystyle\int_{\pi/2}^{\pi} \cos \dfrac{\phi}{2} \, d\phi$

20) $\displaystyle\int_{-0.2}^{0.5} (1 + 0.6 \cos 0.2\theta) \, d\theta$

21) $\displaystyle\int_0^{\pi} (\sin x - \sin 3x) \, dx$

22) $\displaystyle\int_0^1 e^x \, dx$

23) $\displaystyle\int_1^2 e^{-2t} \, dt$

24) $\displaystyle\int_{0.5}^1 \left(e^{\theta/3} - \dfrac{1}{e^{\theta/3}} \right) d\theta$

25) $\displaystyle\int_1^2 (1 + 2e^{0.3v}) \, dv$

26) $\displaystyle\int_2^3 4(e^u + e^{-u}) \, du$

27) $\displaystyle\int_2^3 \dfrac{2}{x} \, dx$

INTEGRATION BY SUBSTITUTION

Previously we learnt how to differentiate by using a substitution and thus reduce a complicated looking expression to one which was relatively simple.

Integration may also be simplified by using a suitable substitution, as shown in the following examples.

EXAMPLE 15.7

Find $\int \left(\dfrac{x}{1 + x^2} \right) dx$

Choosing a suitable substitution is a question of experience, but after working through a number of problems you will find it is not as difficult as it may first appear.

In this instance we will try to simplify the bottom line by putting $u = 1 + x^2$

Hence $\int \left(\dfrac{x}{1 + x^2} \right) dx = \int \left(\dfrac{x}{u} \right) dx$ where $u = 1 + x^2$

$\therefore \qquad \dfrac{du}{dx} = 2x$

$\therefore \qquad dx = \dfrac{du}{2x}$

At this stage we have two variables, x and u, and we shall try to eliminate all the x terms, starting with the dx term.

Thus $\int \left(\dfrac{x}{1 + x^2} \right) dx = \int \left(\dfrac{x}{u} \right) \left(\dfrac{du}{2x} \right)$

$= \tfrac{1}{2} \int \dfrac{1}{u} du$ \qquad which is a standard integral.

$= \tfrac{1}{2} (\log_e u) + c$ \qquad note the introduction of the constant c, as this is an indefinite integral.

At this stage all that remains is to express the variable u in terms of x.

Thus $\int \left(\dfrac{x}{1 + x^2} \right) dx = \tfrac{1}{2} \log_e (1 + x^2) + c$

EXAMPLE 15.8

Find $\int\left(\dfrac{x}{\sqrt{a^2 + x^2}}\right) dx$

Again we shall try a substitution to simplify the bottom line, putting $u = a^2 + x^2$:

$$\int\left(\frac{x}{\sqrt{a^2 + x^2}}\right) dx = \int \frac{x}{\sqrt{u}}\, dx \qquad \text{where} \qquad u = a^2 + x^2$$

$$= \int\left(\frac{x}{u^{1/2}}\right)\left(\frac{du}{2x}\right) \quad \therefore \qquad \frac{du}{dx} = 2x$$

$$= \tfrac{1}{2}\int u^{-1/2}\, du \quad \therefore \qquad dx = \frac{du}{2x}$$

$$= \tfrac{1}{2}\left(\frac{u^{1/2}}{\frac{1}{2}}\right) + c$$

$$= (a^2 + x^2)^{1/2} + c$$

CHANGE OF LIMITS

Definite integrals may also be evaluated using a substitution and changing the limits to values of the new variables. This avoids the necessity of reintroducing the original variable.

EXAMPLE 15.9

Evaluate $\displaystyle\int_0^1 \left(\frac{1}{(2x + 3)^3}\right) dx$

This is an integral with respect to variable x, which means that the limits are also values of x. We shall try the substitution $u = 2x + 3$ and this relationship may be used to find values of u which correspond to the limit values of $x = 1$ and $x = 0$.

$$\int_0^1 \left(\frac{1}{(2x+3)^3} \right) dx = \int_3^5 \left(\frac{1}{u^3} \right) \left(\frac{du}{2} \right)$$

where $u = 2x + 3$

$$= \frac{1}{2} \int_3^5 (u^{-3}) \, du$$

$\therefore \quad \dfrac{du}{dx} = 2$

$$= \frac{1}{2} \left[\frac{u^{-2}}{-2} \right]_3^5$$

$\therefore \quad dx = \dfrac{du}{2}$

$$= \left(\frac{1}{2} \right) \left(\frac{1}{-2} \right) \left[\frac{1}{u^2} \right]_3^5$$

Change of limits:

When $x = 1 \quad u = 2(1) + 3 = 5$

$$= -\frac{1}{4} \left(\frac{1}{5^2} - \frac{1}{3^2} \right)$$

When $x = 0 \quad u = 2(0) + 3 = 3$

$$= 0.0178$$

EXAMPLE 15.10

Evaluate $\displaystyle \int_0^{\pi/2} \sin^3 \theta \cos \theta \, d\theta$

Here the most complicated expression is $\sin^3 \theta$ or $(\sin \theta)^3$ and we will try putting $u = \sin \theta$, and also change the limits.

$$\int_0^{\pi/2} (\sin \theta)^3 (\cos \theta) \, d\theta = \int_0^1 u^3 (\cos \theta) \left(\frac{du}{\cos \theta} \right)$$

where $u = \sin \theta$

$$= \int_0^1 u^3 \, du$$

$\therefore \quad \dfrac{du}{d\theta} = \cos \theta$

$$= \left[\frac{u^4}{4} \right]_0^1$$

$\therefore \quad d\theta = \dfrac{du}{\cos \theta}$

Limits change:

$$= \frac{1}{4}(1^4 - 0^4)$$

when $\theta = \dfrac{\pi}{2}$

$$= 0.25$$

$$u = \sin \frac{\pi}{2} = 1$$

and $\theta = 0$

$$u = \sin 0 = 0$$

Exercise 15.2

By using suitable substitutions find the following indefinite integrals and evaluate the definite integrals:

1) $\int (2x + 1)^3 \, dx$

2) $\int \sqrt{(x + 3)} \, dx$

3) $\int \dfrac{1}{(1 - 2x)} \, dx$

4) $\int x(3 - x^2)^3 \, dx$

5) $\int \sin(2\theta - 1) \, d\theta$

6) $\int \dfrac{dx}{e^{(2x - 1)}}$

7) $\int_4^5 \dfrac{dx}{(x - 3)}$

8) $\int_2^3 x^2(\sqrt{x^3 + 2}) \, dx$

9) $\int_1^2 e^{(t - 4)} \, dt$

10) $\int_0^{\pi/2} \cos\left(3\theta - \dfrac{\pi}{2}\right) d\theta$

11) $\int_2^3 x \, e^{x^2} \, dx$

12) $\int_3^4 \dfrac{1}{x} \log_e x \, dx$

13) $\int_0^1 \dfrac{dx}{(3x + 1)^2}$

14) $\int_0^1 x(\sqrt{1 - x^2}) \, dx$

15) $\int_3^4 \dfrac{x \, dx}{\sqrt[3]{x^2 - 7}}$

16) $\int_0^\pi \sin\theta \cos^2\theta \, d\theta$

17) $\int_{-2}^{-1} \dfrac{e^t}{1 - e^t} \, dt$

18) $\int_0^\pi x \cos(x^2) \, dx$

19) $\int_0^{\pi/4} \dfrac{\cos\theta}{1 - \sin\theta} \, d\theta$

20) $\int_0^2 \dfrac{x^2 \, dx}{\sqrt{x^3 + 1}}$

21) $\int_1^2 \dfrac{(x + 1)}{x^2 + 2x + 2} \, dx$

INTEGRATION BY PARTS

We often need to integrate the product of two functions, for example $\int xe^{2x}\,dx$, and so we shall find a rule to cover such cases.

Now for differentiating products we have

if
$$y = u \times v$$

then
$$\frac{dy}{dx} = v\frac{du}{dx} + u\frac{dv}{dx}$$

and if we integrate both sides with respect to x then

$$\int \frac{dy}{dx}\,dx = \int v\frac{du}{dx}\,dx + \int u\frac{dv}{dx}\,dx$$

or simplifying,
$$\int 1\,dy = \int v\,du + \int u\,dv$$

giving
$$y = \int v\,du + \int u\,dv$$

But $y = uv$, so
$$uv = \int v\,du + \int u\,dv$$

and rearranging
$$\boxed{\int u\,dv = uv - \int v\,du}$$

EXAMPLE 15.11

Find $\int xe^{2x}\,dx$

If we compare $\int xe^{2x}\,dx$ with the LHS of the rule $\int u\,dv$ we may put

$$u = x \quad \text{and} \quad dv = e^{2x}\,dx$$

Thus
$$\frac{du}{dx} = 1 \quad \text{and} \quad \int 1\,dv = \int e^{2x}\,dx$$

giving
$$du = dx \quad \text{and} \quad v = \tfrac{1}{2}e^{2x}$$

If we substitute these in the established rule

$$\int u \, dv \;=\; uv - \int v \, du$$

Then
$$\int x(e^{2x} \, dx) \;=\; x(\tfrac{1}{2} e^{2x}) - \int (\tfrac{1}{2} e^{2x}) \, dx$$

$$=\; \tfrac{1}{2} x e^{2x} - \tfrac{1}{2} \int e^{2x} \, dx$$

Hence
$$\int x e^{2x} \, dx \;=\; \tfrac{1}{2} x e^{2x} - \tfrac{1}{4} e^{2x} + c$$

As for the solution of all indefinite integrals we introduce one constant c after the final integration.

Beware! Suppose that the same integral has been given to you as $\int e^{2x} x \, dx$ and that, using the procedure followed previously,

comparing $\int e^{2x} x \, dx$ with the LHS of the rule $\int u \, dv$

gives
$$u = e^{2x} \qquad \text{and} \qquad dv = x \, dx$$

Then
$$\frac{du}{dx} = 2e^{2x} \qquad \text{and} \qquad \int 1 \, dv = \int x \, dx$$

giving
$$du = 2e^{2x} \, dx \quad \text{and} \qquad v = \frac{x^2}{2}$$

and if we substitute these in the established rule

$$\int u \, dv \;=\; uv - \int v \, du$$

then
$$\int e^{2x} (x \, dx) \;=\; (e^{2x})\left(\frac{x^2}{2}\right) - \int \frac{x^2}{2} 2e^{2x} \, dx$$

All is satisfactory except for the last term – this integral is not only a product but is more complicated than the original problem. So this choice for u and dv will not work. Thus if you had been given the problem as $\int e^{2x} x \, dx$ it would first have been necessary to change the order to $\int x e^{2x} \, dx$.

To avoid this disappointment two hints should be remembered:

(1) Always choose the 'u' function so that it will become simpler on differentiation.
(2) Always choose the 'dv' function that can be integrated easily.

You will not always be right first time but experience helps!

EXAMPLE 15.12

Find $\displaystyle\int_{0.5}^{1} x^2 \log_e x \, dx$

Comparing this with $\displaystyle\int u \, dv$ we could put $u = x^2$ and $dv = \log_e x \, dx$

Now the u term will become simpler on differentiation, but what about integrating the expression for dv? – horrible, to say the least. So we shall try writing the given product the other way round in the integral.

The problem is now $\displaystyle\int_{0.5}^{1} (\log_e x) x^2 \, dx$

It is easier to forget the limits until the integration has been completed and then reintroduce them.

Comparing this with $\displaystyle\int u \, dv$ we may put

$$u = \log_e x \quad \text{and} \quad dv = x^2 \, dx$$

thus
$$\frac{du}{dx} = \frac{1}{x} \quad \text{and} \quad \int 1 \, dv = \int x^2 \, dx$$

giving
$$du = \frac{1}{x} \, dx \quad \text{and} \quad v = \frac{x^3}{3}$$

Substituting these in the rule

$$\int u \, dv = uv - \int v \, du$$

Then
$$\int (\log_e x)(x^2 \, dx) = (\log_e x)\left(\frac{x^3}{3}\right) - \int \frac{x^3}{3}\left(\frac{1}{x} \, dx\right)$$

$$= \tfrac{1}{3} x^3 \log_e x - \tfrac{1}{3} \int x^2 \, dx$$

$$= \tfrac{1}{3} x^3 \log_e x - \frac{x^3}{9} + c$$

Thus
$$\int_{0.5}^{1} x^2 \log_e x \, dx = \left[\frac{x^3}{3} \log_e x - \frac{x^3}{9}\right]_{0.5}^{1}$$

$$= \left(\frac{1^3}{3} \log_e 1 - \frac{1^3}{9}\right) - \left(\frac{0.5^3}{3} \log_e 0.5 - \frac{0.5^3}{9}\right)$$

$$= -0.0683$$

EXAMPLE 15.13

Find $\int x \sin 2x \, dx$

Comparing this with $\int u \, dv$ we may put

$$u = x \quad \text{and} \quad dv = \sin 2x \, dx$$

Thus

$$\frac{du}{dx} = 1 \quad \text{and} \quad \int 1 \, dv = \int \sin 2x \, dx$$

giving

$$du = dx \quad \text{and} \quad v = -\tfrac{1}{2} \cos 2x$$

Substituting in

$$\int u \, dv = uv - \int v \, du$$

gives

$$\int x \, (\sin 2x \, dx) = x\left(-\frac{1}{2} \cos 2x\right) - \int\left(-\frac{1}{2} \cos 2x\right) dx$$

$$= -\frac{x}{2} \cos 2x + \frac{1}{2} \int \cos 2x \, dx$$

$$= -\frac{x}{2} \cos 2x + \frac{1}{4} \sin 2x + c$$

EXAMPLE 15.14

Find $\int \log_e x \, dx$

We may use integration by parts by rewriting the integral as
$\int (\log_e x)\, 1\, dx$.

Comparing this with $\int u\, dv$ we may put

$$u = \log_e x \quad \text{and} \quad dv = 1\, dx$$

Thus
$$\frac{du}{dx} = \frac{1}{x} \quad \text{and} \quad \int 1\, dv = \int 1\, dx$$

giving
$$du = \frac{1}{x}\, dx \quad \text{and} \quad v = x$$

Substituting in
$$\int u\, dv = uv - \int v\, du$$

gives
$$\int (\log_e x)(1\, dx) = (\log_e x)\, x - \int x \left(\frac{1}{x}\, dx \right)$$

$$= (\log_e x)\, x - \int 1\ dx$$

$$= x \log_e x - x + c$$

Exercise 15.3

Using integration by parts, find the following indefinite integrals and evaluate the definite integrals:

1) $\int x e^x\, dx$

2) $\int 3x(\log_e 5x)\, dx$

3) $\int 2x\, e^{3x}\, dx$

4) $\int_0^{\pi/2} x\, \cos x\, dx$

5) $\int_{\pi/2}^{\pi} x \sin x\, dx$

6) $\int_1^2 \log_e x\, dx$

7) $\int_0^1 x e^{-x}\, dx$

8) $\int_0^{\pi} (\pi - x) \cos\ x\ dx$

9) $\int_1^2 \sqrt{x} \log_e x\, dx$

Applications of Integration

16

Areas under graphs – use of areas in practical problems e.g. work done – mean and root mean square (r.m.s) values.

INTRODUCTION

The application of integration to the finding of areas under graphs is extremely important in technology.

You may already have met some instances in engineering where it was necessary to calculate areas under graphs, and we had to use area rules which gave approximate results.

Now, providing we have the equation of a graph, we may find the area underneath the graph using integration and obtain a much more accurate value, and in a much shorter time.

AREA UNDER A GRAPH

Suppose that we wish to find the shaded area shown in Fig. 16.1. P is a point on the curve whose coordinates are (x, y).

Let us now draw, below P, a vertical strip whose width δx is very small. Since the width of the strip is very small we may consider the strip to be a rectangle with height y. Hence the area of the strip is approximately $(y \times \delta x)$. Such a strip is called an elementary strip and we will consider that the shaded area is made up from many elementary strips. Hence the required area is the sum of all the elementary strip areas between the values $x = a$ and $x = b$.

In mathematical notation this may be stated as

$$\text{Area} = \sum_{x=a}^{x=b} y \, \delta x \quad \text{approximately}$$

The process of integration may be considered to sum up an infinite number of elementary strips and hence gives an exact result.

$$\boxed{\text{Area} = \int_a^b y \, dx \ \text{exactly}}$$

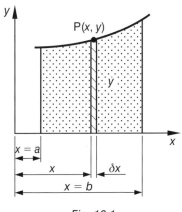

Fig. 16.1

EXAMPLE 16.1

Find the area bounded by the curve $y = x^3 + 3$, the x-axis and the lines $x = 1$ and $x = 3$.

It is always wish to sketch the graph of the given curve and show the area required together with an elementary strip, as shown in Fig. 16.2.

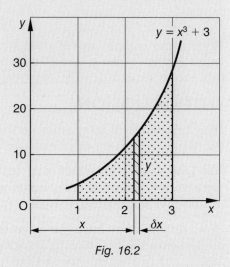

Fig. 16.2

The required area $= \displaystyle\sum_{x=1}^{x=3} y \times \delta x$ approximately

$$= \int_1^3 y \, dx \ \text{exactly}$$

$$= \int_1^3 (x^3 + 3) \, dx$$

$$= \left[\frac{x^4}{4} + 3x \right]_1^3 = \left(\frac{3^4}{4} + 3 \times 3 \right) - \left(\frac{1^4}{4} + 3 \times 1 \right)$$

$$= 26 \text{ square units}$$

EXAMPLE 16.2

Find the area under the curve of $2 \cos \theta$ between $\theta = 20°$ and $\theta = 60°$.

The curve of $2 \cos \theta$ is shown
in Fig. 16.3 and the required area
together with
an elementary strip.

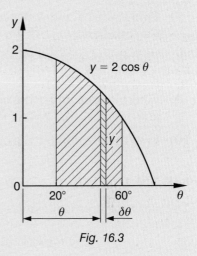

Fig. 16.3

The required area $= \displaystyle\sum_{\theta = 20°}^{\theta = 60°} y \times \delta\theta$ approximately

$$= 2 \int_{20°}^{60°} \cos \theta \, d\theta \quad \text{exactly}$$

$$= 2 \Big[\sin \theta \Big]_{20°}^{60°} \quad = \quad 2(\sin 60° - \sin 20°)$$

$$= 1.05 \text{ square units}$$

Exercise 16.1

1) Find the area between the curve $y = x^3$, the x-axis and the lines
   $x = 5$ and $x = 3$.

2) Find the area between the curve $y = 3 + 2x + 3x^2$, the x-axis and
   the lines $x = 1$ and $x = 4$.

3) Find the area between the curve $y = x^2(2x - 1)$, the x-axis and the
   lines $x = 1$ and $x = 2$.

4) Find the area between the curve $y = \dfrac{1}{x^2}$, the x-axis and the lines
   $x = 1$ and $x = 3$.

5) Find the area between the curve $y = 5x - x^3$, the x-axis and the
   lines $x = 1$ and $x = 2$.

6) Evaluate the integral $\int_0^{2\pi} \sin\theta\, d\theta$ and explain the result with reference to a sketched graph.

7) Find the area under the curve $2\sin\theta + 3\cos\theta$ between $\theta = 0$ and $\theta = \pi$ radians.

8) Find the area under the curve of $y = \sin\phi$ between $\phi = 0$ and $\phi = \pi$ radians.

9) Find the area under the curve $y = e^x$ between the coordinates $x = 0$ and $x = 2$.

10) Find the area under the curve $y = 5e^x$ from $x = -0.5$ to $x = +0.5$

MEAN VALUE

The mean (or average) value or height of a curve is often of importance.

$$\text{The mean value} \ = \ \frac{\text{Area under the curve}}{\text{Length of the base}}$$

A type of graph which is met frequently in technology is a waveform.

A waveform is a graph which repeats indefinitely. A sine curve (Fig. 16.4) is an example of a waveform.

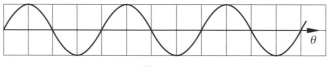

Fig. 16.4

A portion of the graph which shows the complete shape of the waveform without any repetition is called a **cycle**.

In the case of the curve of $\sin\theta$ the portion of the graph over one cycle is said to be a **full wave** (Fig. 16.5), and over half of one cycle is said to be a **half wave** (Fig. 16.6).

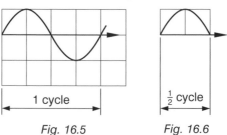

| 1 cycle | $\frac{1}{2}$ cycle |

Fig. 16.5   *Fig. 16.6*

EXAMPLE 16.3

Find the mean value of $A \sin \theta$ for:

a) a half wave   **b)** a full wave

a)

A half wave of $A \sin \theta$
occurs over a range
from $\theta = 0$ rad to $\theta = \pi$ rad
as shown in Fig. 16.7.

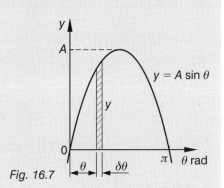

Fig. 16.7

Area of elementary strip $= y \delta \theta$

∴   Total area under curve $= \displaystyle\sum_{\theta = 0}^{\theta = \pi} y \delta \theta$ approximately

$$= \int_0^\pi y \, d\theta \quad \text{exactly}$$

$$= \int_0^\pi A \sin \theta \, d\theta \;=\; A \int_0^\pi \sin \theta \, d\theta$$

$$= A\Big[-\cos \theta \Big]_0^\pi \;=\; A\{(-\cos \pi) - (-\cos 0)\}$$

$$= A\{-(-1) - (-1)\} \;=\; 2A \text{ square units}$$

Now   Mean value $= \dfrac{\text{Area under curve}}{\text{Length of base}} = \dfrac{2A}{\pi} = 0.637A$

b)

A full wave of $A \sin \theta$
occurs over a range
from $\theta = 0$ rad to
$\theta = 2\pi$ rad
as shown in Fig. 16.8.

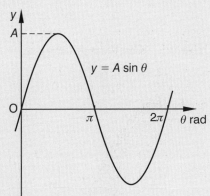

Fig. 16.8

Total area under curve $= \displaystyle\int_0^{2\pi} A \sin \theta \, d\theta \;=\; A\Big[-\cos \theta \Big]_0^{2\pi} \;=\; 0$

This answer is zero because the area under the second half of the wave is
calculated as negative and added to an identical positive area under the
first half wave. Thus the mean value is zero.

EXAMPLE 16.4

A ramp waveform consists of a series of right-angled triangles as shown in Fig. 16.9. Find the mean height of the waveform.

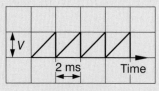

Fig. 16.9

The triangular area under the graph is $\frac{1}{2} \times 2 \times V = V$. However, it is instructive to find the area by integration, as a similar method is used later for finding root mean squares.

We must first find an equation for the sloping side of the triangle, which is shown set up on v–t axes as shown in Fig. 16.10.

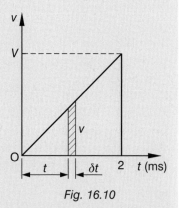

The equation is

$$v = (\text{gradient})t$$

or

$$v = \left(\frac{V}{2}\right)t$$

Fig. 16.10

Now   The triangular area = Sum of elementary strip areas

$$= \sum_{t=0}^{t=2} v\delta t \text{ approximately}$$

$$= \int_0^2 v\,dt \text{ exactly}$$

$$= \int_0^2 \left(\frac{V}{2}\right)t\,dt$$

$$= \frac{V}{2}\int_0^2 t\,dt = \frac{V}{2}\left[\frac{t^2}{2}\right]_0^2 = V$$

Now   Mean value $= \dfrac{\text{Area under graph}}{\text{Length of base}}$

$$= \frac{V}{2}$$

ROOT MEAN SQUARE (r.m.s.) VALUE

In alternating current work the mean value is not of great importance, because we are usually interested in the power produced and this depends on the square of the current or voltage values. In these cases we use the **root mean square** (r.m.s.) value.

The root mean square value $= \sqrt{\text{Average height of the } y^2 \text{ curve}}$

i.e. $\text{r.m.s.} = \sqrt{\dfrac{\text{Area under the } y^2 \text{ curve}}{\text{Length of the base}}}$

EXAMPLE 16.5

Find the r.m.s. value of $A \sin \theta$ for:

a) a half wave   **b)** a full wave

a) The method is similar to that used in Example 16.3 except that we use y^2 instead of y in the integral.

Thus

Total area under y^2 curve $= \displaystyle\int_0^\pi y^2 \, d\theta = \int_0^\pi (A \sin \theta)^2 \, d\theta = A^2 \int_0^\pi \sin^2 \theta \, d\theta$

Now remember how we used the trig, identity $\tan \theta = \dfrac{\sin \theta}{\cos \theta}$ in order to differentiate $\tan \theta$ by use of the quotient formula? Well, in order to integrate $\sin^2 \theta$ we have to use an identity which you have not met before: $\sin^2 \theta = \dfrac{1}{2}(1 - \cos 2\theta)$. This will not cause us any difficulty!!

$\therefore$ Total area under y^2 curve $= \dfrac{A^2}{2} \displaystyle\int_0^\pi (1 - \cos 2\theta) \, d\theta$

$= \dfrac{A^2}{2} \left[\theta - \tfrac{1}{2} \sin 2\theta \right]_0^\pi$

$= \dfrac{A^2}{2} \{ (\pi - \tfrac{1}{2} \sin 2 \times \pi) - (0 - \tfrac{1}{2} \sin 2 \times 0) \}$

$= \dfrac{A^2}{2} \{ \pi - \tfrac{1}{2} \times 0 - 0 + \tfrac{1}{2} \times 0 \}$

$= \dfrac{\pi A^2}{2}$

Now   r.m.s. value $= \sqrt{\dfrac{\text{Area under } y^2 \text{ curve}}{\text{Length of base}}}$

$= \sqrt{\dfrac{\pi A^2/2}{\pi}}$

$= \dfrac{A}{\sqrt{2}} = 0.707A$

b) For a full wave the working is similar to that in part a) except that the limits are 0 to 2π rad. The reader may find it useful to verify that the same result of $0.707A$ is obtained.

Similar results are also obtained for cosine waveforms. Thus:

> For sinusoidal waveforms the r.m.s. value is 0.707 of the amplitude or peak value.

EXAMPLE 16.6

Find the r.m.s. value of the ramp waveform shown in Fig. 16.9.

The method is similar to that used in Example 16.4 except that we use v^2 instead of v in the integral.

Thus

Total area under the v^2 graph $= \displaystyle\int_0^2 v^2 \, dt$

$= \displaystyle\int_0^2 \left\{ \left(\dfrac{V}{2}\right) t \right\}^2 dt$

$= \dfrac{V^2}{4} \displaystyle\int_0^2 t^2 \, dt$

$= \dfrac{V^2}{4} \left[\dfrac{t^3}{3} \right]_0^2 = \dfrac{V^2}{4} \left(\dfrac{2^3}{3} - \dfrac{0^3}{3} \right)$

$= \dfrac{2V^2}{3}$

Now   r.m.s. value $= \sqrt{\dfrac{\text{Area under } v^2 \text{ graph}}{\text{Length of base}}}$

$= \sqrt{\dfrac{2V^2/3}{2}}$

$= \dfrac{V}{\sqrt{3}} = 0.577V$

Exercise 16.2

1) Find the mean value of $V \cos \theta$ for: **a)** a half cycle   **b)** a full cycle

2) Find the mean value of $6 \sin 2\theta$ for: **a)** a half cycle   **b)** a full wave

Find the mean values of the waveforms shown in questions **3–6** over one cycle in each case.

3)
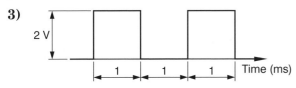

4)
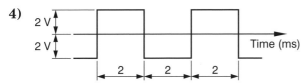

5)
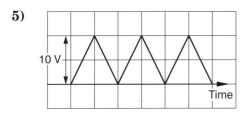

6)
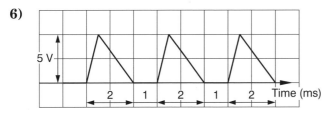

7) Find the mean value of the waveform whose shape for one cycle is as shown.
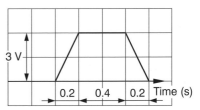

8) Find the r.m.s. value of $V \cos \theta$ over: **a)** a half wave   **b)** a full cycle

Hint: you will need to use the identity $\cos^2 \theta = \dfrac{1}{2}(1 + \cos 2\theta)$

9) Find the r.m.s. value of $6 \sin 2\theta$ over: **a)** a half cycle   **b)** a full cycle

Hint: you will need to use the identity $\sin^2 2\theta = \dfrac{1}{2}(1 - \cos 4\theta)$

10) Find the r.m.s. values of the waveforms in questions **3–7** over one cycle in each case.

17 Areas, Volumes and Centroids

Idea of a centroid – first moment of area – centroid of a composite area – areas and their centroids by integration – volumes of revolution – centroids of volumes by integration.

INTRODUCTION

You will know from everyday observation how important it is to know the positions of centres of gravity – in the design of motor vehicles or aircraft for instance.

Anything which has mass has a centre of gravity – in a similar manner an area or a volume has a centroid.

Here we use the tool called 'calculus' ! We use the idea of integration to sum up elementary components in our quest to determine areas, volumes and the position of their centroids.

Computer programs are available to find centroids but, to enable the user to appreciate how they are applied, it is necessary to have a basic understanding of the topic.

CENTROIDS

The centroid of an area is the point at which the total area may be considered to be situated for calculation purposes. It corresponds to the centre of gravity of a lamina of similar shape to the area. A thin flat sheet of steel of uniform thickness is an example of a lamina.

We know, usually from symmetry, the positions of centres of gravity of common shapes of laminae, such as circular discs and rectangles. Thus we know the positions of the centroid of a circular area and also of a rectangular area.

FIRST MOMENT OF AREA

We know that the moment of a force F about the line AB (Fig. 17.1) is given by $F \times d$.

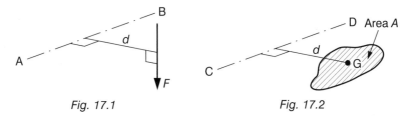

Fig. 17.1   *Fig. 17.2*

Similarly we say that the first moment of the area A about the line CD (Fig. 17.2) is given by $A \times d$, the point G being the centroid of the area.

CENTROIDS OF COMPOSITE AREAS

A composite area is one made up from common shapes.

To determine the position of the centroid G we use two reference axes Ox and Oy as shown in Fig. 17.3.

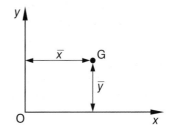

The location of the centroid is given by the co-ordinates $\bar{x}$ and $\bar{y}$.

To find $\bar{x}$ and $\bar{y}$ we use the following formulae, which you may have met previously:

$$\bar{x} = \frac{\Sigma Ax}{\Sigma A} \quad \text{and} \quad \bar{y} = \frac{\Sigma Ay}{\Sigma A}$$

where ΣA is the sum of the component areas

whilst ΣAx and ΣAy are the sum of the first moments of the component areas about the axes Oy and Ox respectively.

EXAMPLE 17.1

Find the position of the centroid shown in Fig. 17.4.

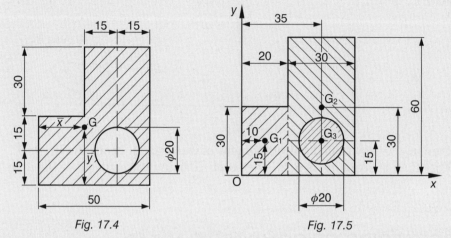

Fig. 17.4   Fig. 17.5

For convenience the reference axes Ox and Oy have been chosen as shown in Fig. 17.5. It will also be seen that the given area has been divided into three component areas whose centroids are G_1, G_2 and G_3.

The dimensions of each component area and the location of each centroid from the axes Ox and Oy are clearly shown. A diagram of this type is essential for this and similar problems – any attempt to obtain these dimensions mentally usually results in one or more errors.

Remember the circle is a 'missing' area and must be taken as negative in the calculations.

It helps to simplify the arithmetic and hence reduce errors by showing the solution in tabular form.

Area	Distance to centroid		Moment of area	
	from Oy	from Ox	about Oy	about Ox
A	x	y	Ax	Ay
$20 \times 30 = \ \ \ 600$	10	15	$600 \times 10 = \ \ \ 6000$	$600 \times 15 = \ \ \ 9000$
$30 \times 60 = 1800$	35	30	$1800 \times 35 = \ \ 63\,000$	$1800 \times 30 = 54\,000$
$-\pi \times 10^2 = -310$	35	15	$-310 \times 35 = -10\,900$	$-310 \times 15 = -4700$
$\Sigma A = \ 2090$			$\Sigma Ax = \ \ 58\,100$	$\Sigma Ay = 58\,300$

Hence $\bar{x} = \dfrac{\Sigma Ax}{\Sigma A} = \dfrac{58\,100}{2090} = 27.8\,\text{mm}$

and $\bar{y} = \dfrac{\Sigma Ay}{\Sigma A} = \dfrac{58\,300}{2090} = 27.9\,\text{mm}$

CENTROIDS OF AREAS BY INTEGRATION

There are many shapes which cannot be divided up exactly into either rectangles or circles. These more complicated shapes can be split up into elementary strips which approximate to rectangles. Then by finding the dimensions of each strip etc. the formulae $\bar{x} = \dfrac{\Sigma Ax}{\Sigma A}$ and $\bar{y} = \dfrac{\Sigma Ay}{\Sigma A}$ may be used to find the approximate values of $\bar{x}$ and $\bar{y}$.

However, if we know, or can find, the equation of y in terms of x, by setting up the area on suitable axes, the numerical summing up may be achieved by integration and exact results obtained.

The following examples will explain how integration is used to find a centroid.

EXAMPLE 17.2

Find the position of the centroid of triangle of height H and base length B.

The triangle must be set up on suitable axes. Only experience enables a good choice. In this case it is convenient to turn the triangle through $90°$ so that the apex lies on the vertical y-axis and the base is parallel to it. In this instance the position of the x-axis is unimportant.

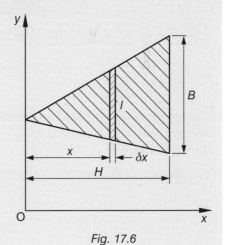

Figure 17.6 shows a sketch of the arrangement and includes the usual elementary strip area of very small width δx and length l.

We shall need a connection between l and x and this may be found by using similar triangles giving

$$\frac{l}{x} = \frac{B}{H} \quad \text{from which} \quad l = \frac{B}{H}x$$

Fig. 17.6

Now the area
of the triangle $= \Sigma A$

$\qquad = $ Sum of the elementary strip areas

$$= \sum_{x=0}^{x=H} l\, \delta x = \int_0^H l\, dx = \int_0^H \frac{B}{H} x\, dx = \frac{B}{H} \int_0^H x\, \delta x$$

$$= \frac{B}{H} \left[\frac{x^2}{2} \right]_0^H = \frac{B}{H} \left(\frac{H^2}{2} - \frac{0^2}{2} \right) = \tfrac{1}{2} BH$$

Also the first moment of the triangular area about the y-axis is

$\qquad \Sigma Ax = $ Sum of the first moment of area of each of the
$\qquad\qquad\qquad$ elementary strips about Oy

$\qquad\qquad = $ Sum of (area of strip $\times$ distance of its centroid from the
$\qquad\qquad\qquad\qquad\qquad\qquad\qquad\qquad\qquad\qquad\qquad\qquad$ y-axis)

$$= \sum_{x=0}^{x=H} l\, \delta x\, x = \int_0^H l\, dx\, x = \int_0^H \frac{B}{H} x\, dx\, x = \frac{B}{H} \int_0^H x^2\, dx$$

$$= \frac{B}{H} \left[\frac{x^3}{3} \right]_0^H = \frac{B}{H} \left(\frac{H^3}{3} - \frac{0^3}{3} \right) = \tfrac{1}{3} BH^2$$

Hence $\qquad \bar{x} = \dfrac{\Sigma Ax}{\Sigma A} = \dfrac{\tfrac{1}{3} BH^2}{\tfrac{1}{2} BH} = \tfrac{2}{3} H$

It should be noted that this is independent of the base length B.

You may remember that the position of
the centroid of a triangular area is as
shown in Fig. 17.7.

The above calculations verify this.

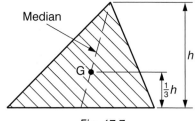

Fig. 17.7

EXAMPLE 17.3

Find the position of the centroid of the sector of a circle of radius R and angle $2a$.

The sector area is shown in Fig. 17.8 located on suitable axes Ox and Oy. By symmetry the centroid G of the sector area lies on the x-axis, and we need to find the distance $\bar{x}$. It is not convenient to divide the area into vertical (or even horizontal) elementary strip areas. Instead we consider elementary sector areas of small angle $\delta\theta$ (Fig. 17.9). Each of these elementary areas may be considered to approximate to a triangle having base length $R \times \delta\theta$ (since length of arc = angle $\times$ radius, where the angle is in radians), and height R.

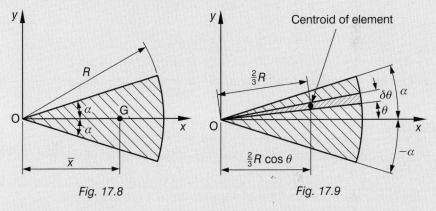

Fig. 17.8   Fig. 17.9

We must first find the area of the sector by summing the areas of the elementary sector areas.

Area of sector $= \Sigma A$

$\qquad = $ Sum of elementary sector areas

$\qquad = \displaystyle\sum_{\theta \,=\, -a}^{\theta \,=\, a} \tfrac{1}{2}(R\,\delta\theta)R \qquad$ since the area of a triangle is $\tfrac{1}{2}$(base)(height)

$\qquad = \displaystyle\int_{-a}^{a} \tfrac{1}{2}(R\,\delta\theta)R \;=\; \tfrac{1}{2}R^2\int_{-a}^{a} d\theta \;=\; \tfrac{1}{2}R^2\big[\theta\big]_{-a}^{a}$

$\qquad = \tfrac{1}{2}R^2\{a - (-a)\} \;=\; R^2 a$

Now the first moment of the given area about the y-axis is

$\Sigma Ax = $ Sum of the first moment of area of each of the elementary sector areas about Oy

$\qquad = $ Sum of (elementary area $\times$ distance of its centroid from the y-axis)

$$= \sum_{\theta=-a}^{\theta=a} (\tfrac{1}{2}R^2\,\delta\theta)(\tfrac{2}{3}R\,\cos\theta) = \int_{-a}^{a} (\tfrac{1}{2}R^2\,d\theta)(\tfrac{2}{3}R\cos\theta)$$

$$= \tfrac{1}{3}R^3 \int_{-a}^{a} \cos\theta\,d\theta = \tfrac{1}{3}R^3\Big[\sin\theta\Big]_{-a}^{a}$$

$$= \tfrac{1}{3}R^3\{\sin a - \sin(-a)\} = \tfrac{1}{3}R^3\{\sin a + \sin a\}$$

$$= \tfrac{2}{3}R^3 \sin a$$

Now $\quad \bar{x} = \dfrac{\Sigma Ax}{\Sigma A} = \dfrac{\tfrac{2}{3}R^3 \sin a}{R^2 a} = \tfrac{2}{3}R\,\dfrac{\sin a}{a}$

It should be remembered that the angle α must be in radians.

EXAMPLE 17.4

Using the result obtained for the position of the centroid of the sector of a circle find:

a) the position of the centroid of a quadrant of a circle of radius R,
b) the position of the centroid of a semicircular area of radius R.

a) If we set the quadrant area on suitable axes as shown in Fig. 17.10, then we may use the formula for the sector of a circle, giving

$$OG = \tfrac{2}{3}R\,\frac{\sin a}{a} \quad \text{where} \quad a = \frac{\pi}{4}\,\text{rad}$$

From right-angled triangle OGM we have

$$\bar{x} = \text{OM} = \text{OG}\left(\cos\frac{\pi}{4}\right)$$

$$= \tfrac{2}{3}R\,\frac{\sin\dfrac{\pi}{4}}{\dfrac{\pi}{4}}\left(\cos\frac{\pi}{4}\right)$$

$$= \frac{2R(0.7071)(0.7071)4}{3\pi}$$

$$= \frac{4R}{3\pi} = 0.424R$$

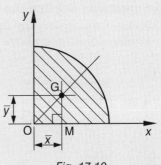

Fig. 17.10

Also by symmetry $\bar{y}$ will have the same value as $\bar{x}$.

b)

The semicircular area shown in Fig. 17.11 may be considered to be two quadrants as indicated. It follows, therefore, that $\bar{x}$ for a semi-circular area is the same as that for a quadrant, i.e. $0.424\,R$.

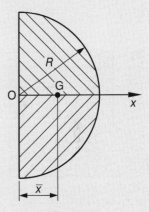

Fig. 17.11

By symmetry the centroid also lies on the horizontal centre line Ox.

Exercise 17.1

In questions **1–4** find the position of the centroids of the cross-sectional areas shown, giving the distances in each case from the left-hand edge (or extreme point) and above the bottom edge (or extreme point).

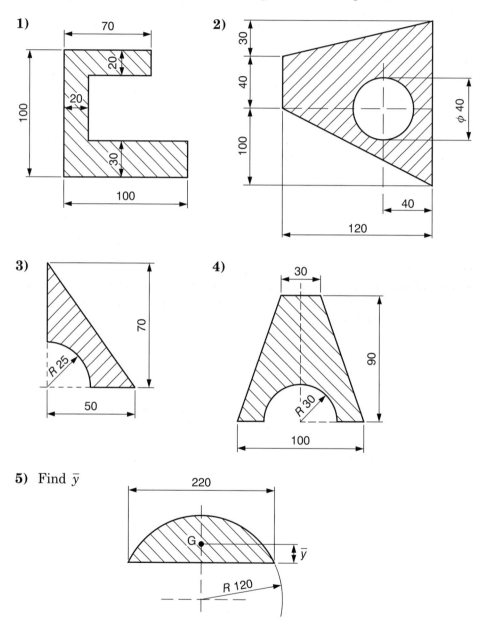

5) Find $\bar{y}$

VOLUMES OF REVOLUTION BY INTEGRATION

If the area under the curve APB (Fig. 17.12) is rotated one complete revolution about the x-axis, then the volume swept out is called a volume of revolution.

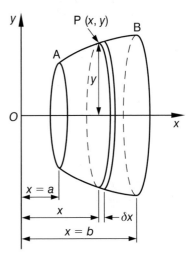

Fig. 17.12

The point P whose co-ordinates are (x, y) is a point on the curve AB.

Let us consider, below P, a thin slice whose width is δx. Since the width of the slice is very small we may consider the slice to be a cylinder of radius y. The volume of this slice is approximately $\pi y^2 \, dx$.

Such a slice is called an elementary slice and we shall consider that the volume of revolution is made up from many such elementary slices. Hence the complete volume of revolution is the sum of all the elementary slices between the values of $x = a$ and $x = b$.

In mathematical notation this may be stated as

$$\sum_{x=a}^{x=b} \pi y^2 \, \delta x \qquad \text{approximately.}$$

As for areas, the process of integration may be considered to sum an infinite number of elementary slices and hence it gives an exact result.

$\therefore \qquad$ Volume of revolution $= \displaystyle\int_a^b \pi y^2 \, dy \qquad$ exactly

EXAMPLE 17.5

The area between the curve $y = x^2$, the x-axis and the ordinates $x = 1$ and $x = 3$ is rotated about the x-axis. Find the volume of revolution.

As when finding areas it is recommended that a sketch is made of the required volume. Figure 17.13 shows a sketch of the required volume.

Required volume

of revolution $= \displaystyle\int_1^3 \pi y^2 \, dx$

$= \displaystyle\int_1^3 \pi (x^2)^2 \, dx$

$= \pi \displaystyle\int_1^3 x^4 \, dx$

$= \pi \left[\dfrac{x^5}{5} \right]_1^3$

$= \pi \left(\dfrac{3^5}{5} - \dfrac{1^5}{5} \right)$

$= 152$ cubic units.

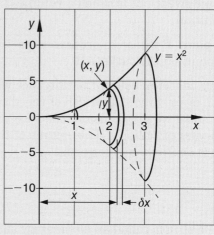

Fig. 17.13

EXAMPLE 17.6

Find, by the calculus, the volume of a cone of base radius R and height H.

The first step is to set up the cone on suitable axes. For convenience the cone has been put with its polar axis lying along the x-axis, as shown in Fig. 17.14.

Then the volume of the cone is the volume of revolution generated when the area OAM is rotated about the x-axis.

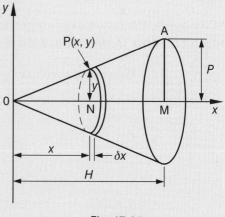

Fig. 17.14

We need an equation connecting x and y. This may be found by considering the similar triangles OPN and OAM.

Then $\quad \dfrac{\text{PN}}{\text{ON}} = \dfrac{\text{AM}}{\text{OM}} \quad$ or $\quad \dfrac{y}{x} = \dfrac{R}{H} \quad$ from which $\quad y = \dfrac{R}{H}\,x$

Now the
$$\text{volume} = \int_0^H \pi y^2\,\mathrm{d}x = \int_0^H \pi\left(\frac{R}{H}x\right)^2 \mathrm{d}x = \pi\frac{R^2}{H^2}\int_0^H x^2\,\mathrm{d}x$$

$$= \pi\frac{R^2}{H^2}\left[\frac{x^3}{3}\right]_0^H = \pi\frac{R^2}{H^2}\left(\frac{H^3}{3} - \frac{0^3}{3}\right)$$

$$= \tfrac{1}{3}\pi R^2 H \quad \text{which verifies a formula you know}$$

EXAMPLE 17.7

Find the volume of a sphere of radius R.

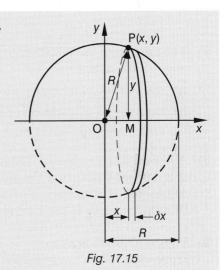

The volume of a sphere may be considered to be a volume of revolution generated by the revolution of a semi-circular area about its boundary diameter.

Suitable axes must now be chosen and these are shown in Fig. 17.15.

Fig. 17.15

The relationship connecting y and x may be found by considering the right-angled triangle OPM and applying the theorem of Pythagoras.

Hence $\quad x^2 + y^2 = R^2 \quad$ from which $\quad y^2 = R^2 - x^2$

Now the
volume of $\quad = \int \pi y^2\,\mathrm{d}x = \pi\int_{-R}^{R}(R^2 - x^2)\,\mathrm{d}x = \pi\left[R^2 x - \frac{x^3}{3}\right]_{-R}^{R}$
revolution

$$= \pi\left\{\left(R^2(R) - \frac{R^3}{3}\right) - \left(R^2(-R) - \frac{(-R)^3}{3}\right)\right\}$$

$$= \tfrac{4}{3}\pi r^3 \quad \text{verifying a formula you should recognise.}$$

Exercise 17.2

In questions **1–5** find the volume generated about the x-axis of the given curves between the limits stated.

1) $y = x^3$ from $x = 0$ to $x = 2$

2) $xy = 16$ from $x = 1$ to $x = 2$

3) $x = 4\sqrt{y}$ from $x = 1$ to $x = 2$

4) $y^3 = x^2$ from $x = 1$ to $x = 8$

5) $y = \dfrac{x}{2}$ from $x = 0$ to $x = 4$

6) Find the volume generated by revolving about the x-axis the portion of the curve $y = x(x - 1)$ that lies below the x-axis.

7) The portion of the circle $x^2 + y^2 = 25$ in the first quadrant is rotated about the x-axis. Find:

a) the volume of a hemisphere of radius 5 m

b) the volume of a section of the above hemisphere cut off by the planes distant 2 m and 3 m from the plane base.

8) A bucket has a radius of 100 mm at the base, and 200 mm at the top. It is 200 mm deep and the sides slope uniformly. Show that it may be considered as a volume of revolution about its axis of symmetry. Hence find the capacity of the bucket, when full, in litres.

9) Figure 17.16 shows the cross-section of a brass nozzle.

Find the volume of the nozzle.

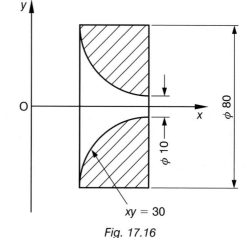

$xy = 30$

Fig. 17.16

CENTROIDS OF VOLUMES BY INTEGRATION

The centroid of a volume is at the point which corresponds to the centre of gravity of a uniform solid body having the same shape. We know the position of the centres of gravity of many solid bodies. For example the centre of gravity of a spherical mass is at its centre; hence we may say that the centroid of a spherical volume is at the centre of the volume.

To find centroids of areas we used the formulae

$$\bar{x} = \frac{\Sigma A x}{\Sigma A} \quad \text{and} \quad \bar{y} = \frac{\Sigma A y}{\Sigma A}$$

For centroids of volumes similar expressions, replacing area A by volume V, are used:

$$\bar{x} = \frac{\Sigma V x}{\Sigma V} \quad \text{and} \quad \bar{y} = \frac{\Sigma V y}{\Sigma V}$$

EXAMPLE 17.8

Find the centroid of a right circular conical volume of height H and base radius R.

For convenience we shall set the cone on axes as shown in Fig. 17.17.

By symmetry the centroid lies on the x-axis.

It remains to find the distance $\bar{x}$ of the centroid from the y-axis.

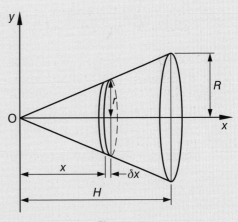

Fig. 17.17

As when finding volumes we consider the cone to be made up from elementary slices, each of radius r and thickness δx, whose volumes are $\pi r^2 \delta x$. The first moment Vx of this slice about the y-axis will be given by $(\pi r^2 \delta x)x$, and if we find the sum of these first moments we obtain the first moment of the whole conical volume. In mathematical notation this may be stated as

Sum of the first moments $\quad \Sigma Vx \;=\; \sum_{x=0}^{x=H} (\pi r^2 \delta x)x \qquad$ approximately

$$= \int_0^H (\pi r^2 \, dx)x \qquad \text{exactly}$$

We must now find r in terms of x. From similar right-angled triangles in Fig. 17.17 we have $\dfrac{r}{R} = \dfrac{x}{H}$ from which $r = \left(\dfrac{R}{H}\right)x$. Thus

$$\Sigma Vx \;=\; \int_0^H \left\{ \pi\left(\frac{R}{H}x\right)^2 dx \right\}x \;=\; \frac{\pi R^2}{H^2} \int_0^H x^3 \, dx \;=\; \frac{\pi R^2}{H^2}\left[\frac{x^4}{4}\right]_0^H$$

$$= \frac{\pi R^2}{H^2} \times \frac{H^4}{4} \;=\; \tfrac{1}{4}\pi R^2 H^2$$

and since we know the volume of a cone is $\tfrac{1}{3}\pi R^2 H$,

$$\bar{x} \;=\; \frac{\Sigma Vx}{\Sigma V} \;=\; \frac{\tfrac{1}{4}\pi R^2 H^2}{\tfrac{1}{3}\pi R^2 H} \;=\; \tfrac{3}{4}H$$

EXAMPLE 17.9

Find position of the centroid of a hemispherical volume of radius 80 mm.

Again, for convenience, we have set the volume on axes as shown in Fig. 17.18.

We consider the volume to comprise elementary slices as shown.

Thus

$$\Sigma Vx \;=\; \sum_{x=0}^{x=80} (\pi y^2 \, \delta x)x \quad \text{approximately}$$

$$= \int_0^{80} (\pi y^2 \, \delta x)x \quad \text{exactly}$$

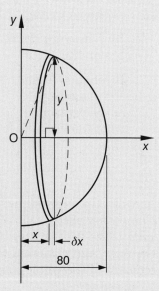

Fig. 17.18

Here we require y in terms of x and apply Pythagoras' theorem to the right-angled triangle in Fig. 17.18 which gives $80^2 = x^2 + y^2$
from which $y^2 = 80^2 - x^2$

Thus $\Sigma Vx = \displaystyle\int_0^{80} \{\pi(80^2 - x^2)\,\mathrm{d}x\}x = \pi\int_0^{80}(80^2x - x^3)\,\mathrm{d}x$

$= \pi\left[\dfrac{80^2x^2}{2} - \dfrac{x^4}{4}\right]_0^{80} = \pi\left(\dfrac{80^2(80)^2}{2} - \dfrac{80^4}{4}\right) = \pi\dfrac{80^4}{4}$

But we know that the volume of a hemisphere of radius 80 is $\frac{2}{3}\pi(80)^3$.

Thus $\bar{x} = \dfrac{\Sigma Vx}{\Sigma V} = \dfrac{\pi\dfrac{80^4}{4}}{\frac{2}{3}\pi(80)^3}$ (Note that, by keeping the powers of 80, the final cancelling avoids much arithmetic!)

$= \frac{3}{8}(80) = 30\,\text{mm}$

Exercise 17.3

Find, using integration, the position of the centroids along the axis of symmetry of the following volumes:

1) A cylinder of length L.

2) A pyramid of height H and square base.

3) A 1.5 m deep frustum of a cone, 1.2 mm diameter at the top and 2.4 m diameter at the bottom.

4) A hemispherical dome with a radius of 4 m and having its top 1 m removed.

5) A 2 m deep cap of a 6 m radius sphere.

18 Second Moment of Area

Idea of a second moment of area – determination by integration – units – conventional notation – composite sections – parallel axis theorem – general procedure for non-symmetrical sectional areas – idea of polar second moment of area – determination by integration – perpendicular axis theorem – radius of gyration.

INTRODUCTION

The second moment of an area is a mathematical property used in calculations. Such calculations are used in designing structural steelwork to find stress and deflections – these will determine the sizes of beams.

The polar second moment of area is a similar property used in the design of components in torsion – the drive shaft from an engine is a good example.

SECOND MOMENT OF AREA

Figure 18.1 shows a thin strip, of area A and of very small width, which is distant x from the reference line CD.

Now the first moment of area A about the reference line CD is given by Ax.

Similarly the second moment of area A about the reference line CD is given by Ax^2.

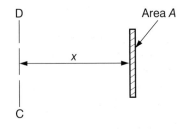

Fig. 18.1

The symbol for the second moment of area is I and is always stated with reference to an axis or datum line. Thus for the strip area shown

$$I_{CD} = Ax^2$$

The expression Ax^2 is only true if the area is a very thin strip parallel to the reference line. It follows that if we require the second moment of any other shaped area it is necessary to divide the area into a number of elementary strips all parallel to the reference line.

Thus in Fig. 18.2 for the irregular area

$$I_{CD} = \Sigma Ax^2$$

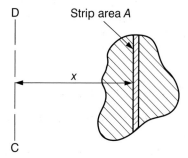

Fig. 18.2

The summation of the Ax^2 for each strip may be carried out by graphical and numerical methods but, as in the case of finding first moments of area, this summation may often be achieved by integration as the following example will illustrate:

EXAMPLE 18.1

Find the second moment of area of a rectangle of width b and depth d about its base edge.

The rectangular area must be set up on suitable axes. In this case it is convenient to turn the rectangle through $90°$ and arrange for the base line to line on the y-axis as shown in Fig. 18.3.

In this instance the position of the x-axis is unimportant.

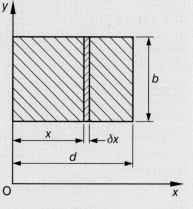

Fig. 18.3

The diagram shows a typical elementary strip area parallel to the reference axis and whose area is $b\,\delta x$.

Now the second moment of the rectangular area about the y-axis

$$= \Sigma A x^2 = \sum_{x=0}^{x=d} (b\,\delta x)x^2 = \int_0^d b\,\mathrm{d}x\,x^2 = b\int_0^d x^2\,\mathrm{d}x$$

$$= b\left[\frac{x^3}{3}\right]_0^d = \frac{bd^3}{3}$$

EXAMPLE 18.2

Find the second moment of a triangular area, of height H and base B, about a line through its apex and parallel to the base.

The arrangement of the axes etc. is identical to that used in Example 17.2 and is as shown in Fig. 18.4.

Now the second moment of the rectangular area about the y-axis

$$= \Sigma A x^2$$

$$= \sum_{x=0}^{x=H} (l\,\delta x)x^2$$

$$= \int_0^H l\,x^2\,\mathrm{d}x$$

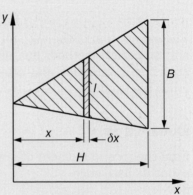

Fig. 18.4

and using similar triangles as before we have $l = \dfrac{B}{H}x$.

Thus
$$I_{\mathrm{OY}} = \int_0^H \left(\frac{B}{H}x\right)x^2\,\mathrm{d}x = \frac{B}{H}\int_0^H x^3\,\mathrm{d}x$$

$$= \frac{B}{H}\left[\frac{x^4}{4}\right]_0^H = \frac{B}{H}\left(\frac{H^4}{4} - \frac{0^4}{4}\right) = \frac{B}{H}\left(\frac{H^4}{4}\right) = \frac{BH^3}{4}$$

UNITS OF SECOND MOMENTS OF AREA

The second moment of area is $\Sigma A x^2$. If all lengths are in metres, then the result will be $m^2 \times m^2 = m^4$.

For example in the formula $\dfrac{bd^3}{3}$ the units are $m \times m^3 = m^4$.

In typical engineering problems it is unlikely that dimensions of a cross-sectional area will be in metres. It is more probable that the units will be millimetres, and if these are used the second moment of area will be in mm^4 which will result in large numbers. In practice it is usual to calculate second moments of area in cm^4. This makes the arithmetic as simple as possible.

This is one of the few times when units not recommended by the Système International (SI) are used in engineering calculations. In British industry producers of rolled steel sections used for beams and columns, etc. agreed with continental manufacturers to use cm^4 as units of second moments of area.

The second moment of the rectangular area shown in Fig. 18.5 about its base, or more correctly about an axis BB on which the base lies, is

$$I_{BB} = \frac{bd^3}{3} = \frac{3 \times 2^3}{3} = 8\ cm^4$$

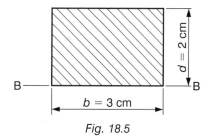

Fig. 18.5

CONVENTIONAL NOTATION I_{XX} AND I_{YY}

XX and YY are known conventionally as the horizontal and vertical axes that pass through the centroid of a cross-sectional area.

Hence: I_{XX} is the second moment of a cross-sectional area about the horizontal axis that passes through the centroid,

and: I_{YY} is the second moment of a cross-sectional area about the vertical axis that passes through the centroid.

EXAMPLE 18.3

Find I_{XX} and I_{YY} of the cross-sectional area shown in Fig. 18.6.

In this case, by symmetry, the centroid G will be at the centre of the section.

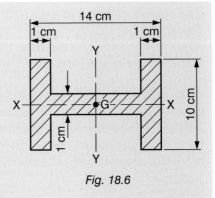

Fig. 18.6

To find I_{XX}

We shall consider the area to be made up of six rectangles, as shown in Fig. 18.7.

Each of these rectangles has its 'base' on XX.

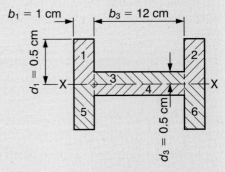

Fig. 18.7

Now I_{XX} for the cross-sectional area will be the sum of the second moments of area of each rectangle about XX.

Rectangles 1, 2, 5 and 6 have the same dimensions and hence the same I_{XX}. Similarly rectangles 3 and 4 have the same I_{XX}.

Hence I_{XX} for the whole area $= 4(I_{XX}$ for area 1$) + 2(I_{XX}$ for area 3$)$

$$= 4\left(\frac{b_1 d_1{}^3}{3}\right) + 2\left(\frac{b_3 d_3{}^3}{3}\right)$$

$$= 4\left(\frac{1 \times 5^3}{3}\right) + 2\left(\frac{12 \times 0.5^3}{3}\right)$$

$$= 167.7 \, \text{cm}^4$$

To find I_{YY}

As before we shall consider the area to comprise six rectangles.

Each rectangle has its 'base' on YY as shown in Fig. 18.8.

In this case the 'bases' are vertical!

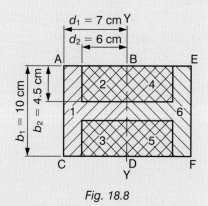

Fig. 18.8

The portion to the left of YY is area 1 (ABCD) less areas 2 and 3. This means that the I_{YY} of both areas 2 and 3 must be taken as negative. We treat the portion the right of YY in a similar manner.

Rectangles 1 and 6 have the same dimensions and hence the same I_{YY}. Similarly rectangles 2, 3, 4 and 5 have the same I_{YY}.

Hence I_{YY} for the whole area $= 2(I_{YY}$ for area 1$) - 4(I_{YY}$ for area 2$)$

$$= 2\left(\frac{b_1 d_1{}^3}{3}\right) - 4\left(\frac{b_2 d_2{}^3}{3}\right)$$

$$= 2\left(\frac{10 \times 7^3}{3}\right) - 4\left(\frac{4.5 \times 6^3}{3}\right)$$

$$= 991 \text{ cm}^4$$

Exercise 18.1

Find I_{XX} and I_{YY} for the cross-sectional areas given in the following examples. All dimensions are in cm.

1)

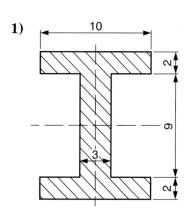

2)

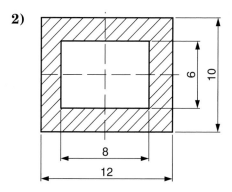

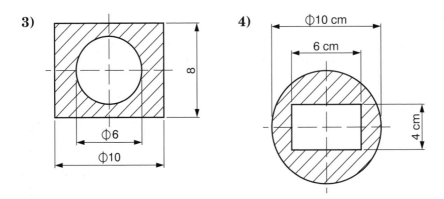

THE PARALLEL AXIS THEOREM

A second moment of area must always be stated together with the axis about which it has been calculated. We state by saying, for example, that I_{AB} is the second moment of area about the axis AB.

Figure 18.9 shows a plane area A whose centroid is G. Also shown is an axis that passes through G and a parallel axis, the distance between the axes being h.

The parallel axis theorem states:

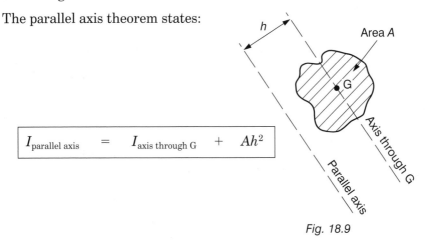

$$I_{\text{parallel axis}} = I_{\text{axis through G}} + Ah^2$$

Fig. 18.9

This is extremely useful as the following examples will show. It is worth while remembering that $I_{\text{axis through G}}$ is the numerically smallest second moment of area.

EXAMPLE 18.4

Find the second moment of area of the rectangle shown in Fig. 18.10 about the axis XX which passes through the centroid.

We know from Example 18.1 that the second moment of area of the rectangle about its base is $\dfrac{bd^3}{3}$.

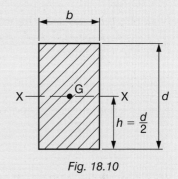

Fig. 18.10

Now using the parallel axis theorem we have

$$I_{\text{base}} = I_{\text{XX}} + Ah^2$$

$\therefore$ rearranging $I_{\text{XX}} = I_{\text{base}} - Ah^2$

$$= \frac{bd^3}{3} - bd\left(\frac{d}{2}\right)^2 = bd^3\left(\frac{1}{3} - \frac{1}{4}\right) = \frac{bd^3}{12}$$

EXAMPLE 18.5

Given that the second moment of a triangular area, of height H and base B, about a line through its apex and parallel to its base is $\dfrac{BH^3}{4}$, find the second moment of the triangular area about its base.

Figure 18.11 shows a sketch of the triangular area with axes suitably labelled. We are given I_{MM} and we require I_{NN}.

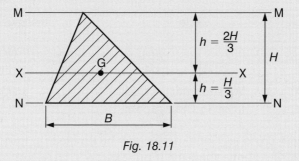

Fig. 18.11

The parallel axis theorem requires that one of the axes used must pass through the centroid. It is necessary, therefore, to use the parallel axis theorem twice – once to find I_{XX} knowing I_{MM} and again to find I_{NN} using I_{XX}.

Now the parallel axis theorem gives $I_{MM} = I_{XX} + Ah_1^2$

and rearranging $I_{XX} = I_{MM} - Ah_1^2$

$\therefore$
$$I_{XX} = \frac{BH^3}{4} - \frac{BH}{2}\left(\frac{2H}{3}\right)^2 = \frac{BH^3}{36}$$

Also the parallel axis theorem gives $I_{NN} = I_{XX} + Ah_2^2$

$$= \frac{BH^3}{36} + \frac{BH}{2}\left(\frac{H}{3}\right)^2 = \frac{BH^3}{12}$$

Hence the second moment of a triangular area about its base is $\dfrac{BH^3}{12}$.

SUMMARY OF SECOND MOMENTS OF COMMON AREAS

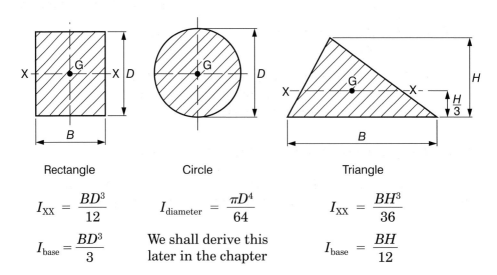

Rectangle

$$I_{XX} = \frac{BD^3}{12}$$

$$I_{base} = \frac{BD^3}{3}$$

Circle

$$I_{diameter} = \frac{\pi D^4}{64}$$

We shall derive this later in the chapter

Triangle

$$I_{XX} = \frac{BH^3}{36}$$

$$I_{base} = \frac{BH}{12}$$

In practice most cross-sectional areas may be divided up into a combination of the above shapes – if not exactly, a close approximation can be achieved.

GENERAL PROCEDURE FOR FINDING I_{XX} AND I_{YY} FOR NON-SYMMETRICAL AREAS

Briefly, the method is to find the second moment of the sectional area about axes OH and OV. We then use the parallel axis theorem to find I_{XX} and I_{YY} which are parallel axes passing through the section centroid G.

Don't be put off by the complicated looking instructions below. Just follow the procedure through in the two examples which follow and you will soon get the idea.

Set up convenient vertical and horizontal axes, OV and OH. Divide the given area into suitable component areas and show them on a new diagram. This should also show the dimensions of each component area and the distances of the respective centroids from axes OV and OH.

To find I_{YY} for the given cross-sectional area:

(1) Find the position of the centroid using $\bar{x} = \dfrac{\Sigma Ax}{\Sigma A}$.

(2) For each component area find:

    (a) I_{YY}^{G} about a vertical axis through its centroid G

    (b) Ax^2 x being the distance between G and OV

    (c) I_{OV} from $I_{OV} = I_{YY}^{G} + Ax^2$,   using the parallel axis theorem.

(3) Find ΣI_{OV} the second moment of the given cross-sectional area about OV.

(4) Find I_{YY} the second moment of the given cross-sectional area about YY, using $I_{YY} = \Sigma I_{OV} - (\Sigma A)(\bar{x})^2$ from the parallel axis theorem.

To find I_{XX} use a sequence similar to the above.

It helps to simplify the arithmetic *and hence reduce errors*, by using a tabular form for solutions as shown in the following examples:

EXAMPLE 18.6

Find I_{XX} and I_{YY} for the cross-sectional area shown in Fig. 18.12. All dimensions are in cm.

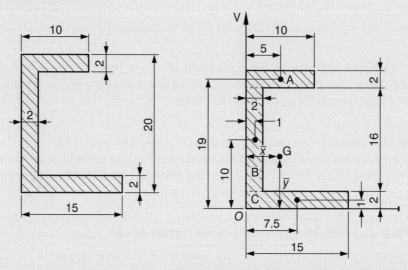

Area	B	D	$A =$ BD	x	Ax	Ax^2	$I_{YY}^G =$ $\dfrac{DB^3}{12}$	$I_{OV} =$ $I_{YY}^G + Ax^2$	y	Ay	Ay^2	$I_{XX}^G =$ $\dfrac{BD^3}{12}$	$I_{OH} =$ $I_{XX}^G + Ay^2$
	cm	cm	cm^2	cm	cm^3	cm^4	cm^4	cm^4	cm	cm^3	cm^4	cm^4	cm^4
A	10	2	20	5	100	500	167	667	19	380	7220	7	7227
B	2	16	32	1	32	32	11	43	10	320	3200	683	3883
C	15	2	30	7.5	225	1688	563	2251	1	30	30	10	40
Totals:			82 ΣA		357 ΣAx			2961 ΣI_{OV}		730 ΣAy			11 150 ΣI_{OH}

$$\bar{x} = \frac{\Sigma Ax}{\Sigma A} = \frac{357}{82} = 4.35 \text{ cm} \qquad \bar{y} = \frac{\Sigma Ay}{\Sigma A} = \frac{730}{82} = 8.90 \text{ cm}$$

$$
\begin{aligned}
I_{YY} &= \Sigma I_{OV} - (\Sigma A)(\bar{x})^2 \\
&= 2961 - (82)(4.35)^2 \\
&= 1410 \text{ cm}^4
\end{aligned}
\qquad
\begin{aligned}
I_{XX} &= \Sigma I_{OH} - (\Sigma A)(\bar{y})^2 \\
&= 11\,150 - (82)(8.90)^2 \\
&= 4650 \text{ cm}^4
\end{aligned}
$$

EXAMPLE 18.7

Find I_{XX} and I_{YY} for the area shown in Fig. 18.13. All dimensions are cm.

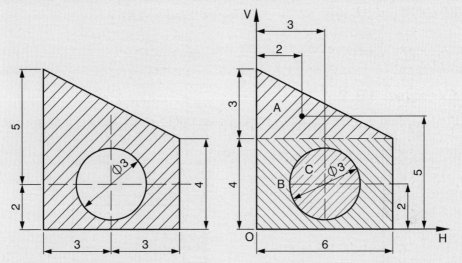

In the table we shall use for:

a rectangle $I_{YY}^G = \dfrac{DB^3}{12}$ and $I_{XX}^G = \dfrac{BD^3}{12}$

a triangle $I_{YY}^G = \dfrac{DB^3}{36}$ and $I_{XX}^G = \dfrac{BD^3}{36}$

a circle $I_{YY}^G = I_{XX}^G = \dfrac{\pi D^4}{64}$ which is taken as negative, since it is a 'missing' area.

Area	B cm	D cm	A cm^2	x cm	Ax cm^3	Ax^2 cm^4	I_{YY}^G cm^4	$I_{OV} =$ $I_{YY}^G + Ax^2$ cm^4	y cm	Ay cm^3	Ay^2 cm^4	I_{XX}^G cm^4	$I_{OH} =$ $I_{XX}^G + Ay^2$ cm^4
A	6	3	9	2	18	36	18	54	5	45	225	5	230
B	6	4	24	3	72	216	72	288	2	48	96	32	128
C	3 diam.		−7.1	3	−21.3	−64	−4	−68	2	−14.2	−28	−4	−32
Totals:			25.9 ΣA		68.7 ΣAx			274 ΣI_{OV}		78.8 ΣAy			326 ΣI_{OH}

$$\bar{x} = \frac{\Sigma Ax}{\Sigma A} = \frac{68.7}{25.9} = 2.65 \text{ cm} \qquad \bar{y} = \frac{\Sigma Ay}{\Sigma A} = \frac{78.8}{25.9} = 3.04 \text{ cm}$$

$$\begin{aligned} I_{YY} &= \Sigma I_{OV} - (\Sigma A)(\bar{x})^2 \\ &= 274 - (25.9)(2.65)^2 \\ &= 92.1 \text{ cm}^4 \end{aligned} \qquad \begin{aligned} I_{XX} &= \Sigma I_{OH} - (\Sigma A)(\bar{y})^2 \\ &= 326 - (25.9)(3.04)^2 \\ &= 86.6 \text{ cm}^4 \end{aligned}$$

Exercise 18.2

Find I_{XX} and I_{YY} for the following cross-sectional areas. Dimensions are cm.

1)

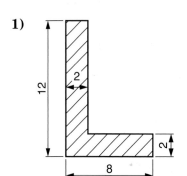

2)

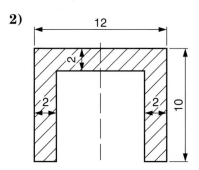

3)

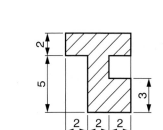

4)

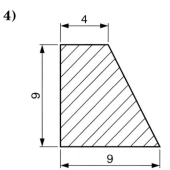

5)

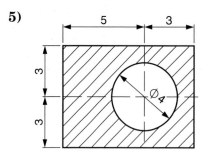

6)

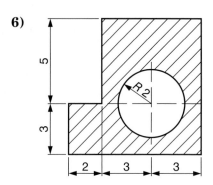

POLAR SECOND MOMENT OF AREA

The polar second moment of area, denoted by the symbol J, is the second moment of a circular area about the polar axis (that is the axis passing through the centre of the area and perpendicular to the plane of the area).

One example of its use is in finding stresses due to torsion in a circular shaft.

Now the second moment of area of the elementary strip area, shown in Fig. 18.14, about the reference line CD is Ax^2.

Similarly we say that the polar second moment of area of the elementary circular strip area, shown in Fig. 18.15, is Ar^2.

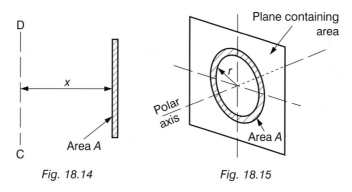

Fig. 18.14   Fig. 18.15

EXAMPLE 18.8

Find the polar second moment of a circular area of diameter D.

Figure 18.16 shows the given area and an elementary circular strip.

We shall sum the Ar^2 for each strip area to find the polar second moment of the whole area.

Hence J for the circular area $= \Sigma Ar^2$.

Now the approximate area A of the elementary circular strip is

circumference of strip $\times$ width of strip $= 2\pi r \times \delta r$

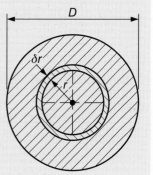

Fig. 18.16

Hence $\quad J = \sum_{r=0}^{r=D/2} (2\pi r \,\delta r)r^2 = \int_0^{D/2} 2\pi r \,dr\, r^2$

$$= 2\pi \int_0^{D/2} r^3 \,dr \quad = 2\pi \left[\frac{r^4}{4}\right]_0^{D/2} = \frac{\pi D^4}{32}$$

EXAMPLE 18.9

Find the polar second moment of area of the cross-section of a tube 6 cm outside diameter and 4 cm inside diameter.

We shall consider J of the cross-sectional area shown in Fig. 18.17 to be J of a 6 cm diameter circular area less J of a 4 cm diameter circular area.

$$\therefore \quad \text{required } J = \frac{\pi D^4}{32} - \frac{\pi d^4}{32}$$

$$= \frac{\pi 6^4}{32} - \frac{\pi 4^4}{32}$$

$$= 127 - 25$$

$$= 102 \text{ cm}^4$$

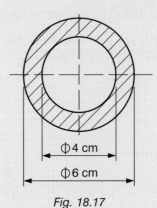

Fig. 18.17

THE PERPENDICULAR AXIS THEOREM

Figure 18.18 shows a plane area on which the axes OX and OY are drawn, the axes being at right angles. OZ is an axis perpendicular to the plane area. OX, OY and OZ are said to be mutually perpendicular.

The perpendicular axis theorem states:

$$\boxed{I_{\text{OZ}} = I_{\text{OX}} + I_{\text{OY}}}$$

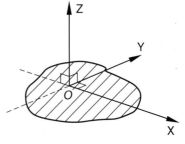

Fig. 18.18

EXAMPLE 18.10

Find I_{OZ} for the rectangle shown in Fig. 18.19.

We know that for a rectangle

$$I_{base} = \frac{bd^3}{3}$$

$$\therefore \quad I_{OA} = \frac{12 \times 8^3}{3} = 2048 \text{ cm}^4$$

and $\quad I_{OB} = \frac{8 \times 12^3}{3} = 4608 \text{ cm}^4$

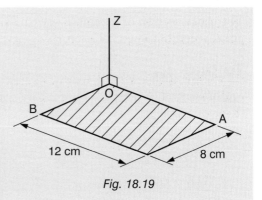

Fig. 18.19

The perpendicular axis theorem gives

$$I_{OZ} = I_{OA} + I_{OB} = 2048 + 4608 = 6656 \text{ cm}^4$$

EXAMPLE 18.11

Using the fact that J for a circular area is $\dfrac{\pi d^4}{32}$ and the perpendicular axis theorem finds I about a diameter.

Now for a circular area I_{OZ} is called J (Fig. 18.20), and both I_{OX} and I_{OY} are equal to $I_{diameter}$.

The perpendicular axis theorem states:

$$I_{OZ} = I_{OX} + I_{OY}$$

$$\therefore \quad J = I_{diameter} + I_{diameter}$$

Thus $\quad I_{diameter} = \tfrac{1}{2}J$

$$= \frac{1}{2} \times \frac{\pi d^4}{32} = \frac{\pi d^4}{64}$$

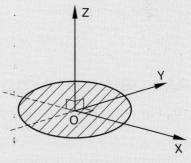

Fig. 18.20

Hence I about a diameter of a circular area is $\dfrac{\pi d^4}{64}$.

A *semicircular area* has I about the boundary diameter equal to one half of the result obtained for a full circular area about a diameter.

Thus for a semicircular area $I_{diameter} = \dfrac{\pi d^4}{128}$

EXAMPLE 18.12

The cross-section of a shaft is shown in Fig. 18.21. Find I_{XX}, I_{YY}, and the polar second moment of area J.

To find I_{XX}

I_{XX} of section

$= I_{XX}$ of square $- I_{XX}$ of circle

$= \dfrac{bd^3}{12} - \dfrac{\pi D^4}{64}$

$= \dfrac{50 \times 50^3}{12} - \dfrac{\pi 30^4}{64}$

$= 520\,800 - 39\,800$

$= 481\,000 \text{ mm}^4$

$= \dfrac{481\,000}{10^4} \text{ cm}^4 = 48.1 \text{ cm}^4$

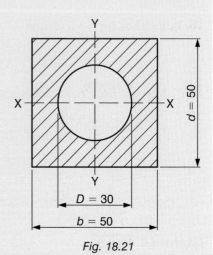

Fig. 18.21

To find I_{YY}

We may see from the symmetry of the figure that $I_{YY} = I_{XX}$.

Hence $\qquad\qquad\qquad I_{YY} = 48.1 \text{ cm}^4$

To find J

Using the perpendicular axis theorem we have:

$$J = I_{XX} + I_{YY} = 48.1 + 48.1 = 96.2 \text{ cm}^4$$

RADIUS OF GYRATION

The formula for finding centroids is

$$(\Sigma A)\bar{x} = \Sigma Ax$$

Similarly for second moments of area we have

$$I = (\Sigma A)k^2 = \Sigma Ax^2$$

where k is called the **radius of gyration** of the area.

In this chapter we have concentrated on using I found from ΣAx^2. However, we are given sometimes the area ΣA of a cross-section together with the radius of gyration k and we then find the second moment of area I by using $\quad I = (\Sigma A)k^2$.

EXAMPLE 18.13

Find I_{XX} for a cross-sectional area given that the area of the section is 85 cm² and $k_{XX} = 3.7$ cm.

We have $\qquad\qquad I = (\text{area of section})k^2$

$\therefore \qquad\qquad I_{XX} = 85(3.7)^2 = 1164 \text{ cm}^4$

Exercise 18.3

1) For the cross-sectional area shown in Fig. 18.22 find:

    a) I about a diameter   **b)** J about the polar axis   **c)** I_{AA}

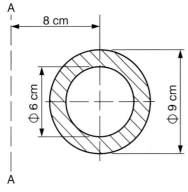

Fig. 18.22

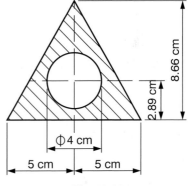

Fig. 18.23

2) For the cross-sectional area shown in Fig. 18.23 find:

    a) I_{XX}, I_{YY} and J **b)** k_{XX}

3) Starting with the knowledge that the second moment of area of a rectangle about an edge is $bd^3/3$ and that the second moment of area of a circle about a diameter is $\pi d^4/64$, and also using the perpendicular and parallel axis theorems, find the second moment of the area shown in Fig. 18.24 about the:

    a) axis A′A

    b) axis through G, perpendicular to the plane of the area

    c) axis B′B

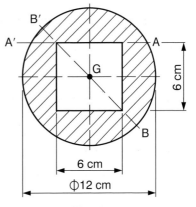

Fig. 18.24

19 *Moment of Inertia*

Moment of inertia I for a rotating mass system – units of I – finding I by integration – parallel axis theorem – I for common shapes.

INTRODUCTION

Moment of inertia is the property of a body used in rotational problems just as the mass is used in problems involving linear motion.

The symbol for moment of inertia is I and must always be stated together with the reference axis about which it has been calculated, in a similar manner as was used for second moment of area. It is unfortunate that engineers have chosen the same symbol for moment of inertia as for second moment of area but, in practice, confusion hardly ever occurs.

I FOR A SYSTEM

The linear kinetic energy of a mass m moving with a velocity v is given by $\frac{1}{2}mv^2$.

Also, the tangential velocity v at the end of a radius r rotating about an axis with angular velocity ω is given by $v = r\omega$.

Consider the system shown in Fig. 19.1 which comprises three concentrated masses m_1, m_2, and m_3, each being fixed to the end of a radius arm. The radius arms are all attached to a vertical spindle, the arms and spindle being assumed to have negligible mass. The whole system rotates with an angular velocity ω.

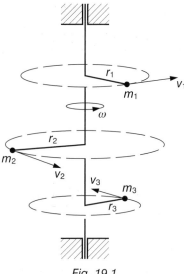

Fig. 19.1

The kinetic energy of the system $= \frac{1}{2}m_1v_1^2 + \frac{1}{2}m_2v_2^2 + \frac{1}{2}m_3v_3^2$

$$= \frac{1}{2}m_1(r_1\omega)^2 + \frac{1}{2}m_2(r_2\omega)^2 + \frac{1}{2}m_3(r_3\omega)^2$$

$$= \frac{1}{2}(m_1r_1^2 + m_2r_2^2 + m_3r_3^2)\omega^2$$

$$= \frac{1}{2}(\Sigma mr^2)\omega^2$$

$$= \frac{1}{2}I\omega^2$$

where the moment of inertia of the system is $I = \Sigma mr^2$.

Although the expression $I = \Sigma mr^2$ has been derived for three rotating masses it is true for a system comprising any number of concentrated masses.

This expression for I may be used, together with integration, for finding the moment of inertia of bodies whose mass is *not* concentrated at any one particular radius. Example 19.3 shows this method.

Now for first moments of area $\qquad (\Sigma A)\bar{x} = (\Sigma Ax)$

and for second moments of area $\qquad I = (\Sigma A)k^2 = (\Sigma Ax^2)$

Similarly for moments of inertia: $\quad \boxed{I = (\Sigma m)k^2 = (\Sigma mr^2)}$

where $\qquad\qquad\qquad \Sigma m$ is the total mass of the body

and $\qquad\qquad\qquad\quad k$ is the radius of gyration

UNITS OF MOMENTS OF INERTIA

Since $I = (\Sigma m)k^2$ the units will be those of mass $\times$ distance^2.
In basic SI units the mass will be kilograms and the distance metres,
and so the units of moment of inertia will be kg m^2.

EXAMPLE 19.1

Find the moment of inertia of a component about an axis, given that its
mass is 5 kg and the radius of gyration about this axis is 200 mm.

Now $\qquad I = \text{Total mass} \times k^2 = 5 \times \left(\dfrac{200}{1000}\right)^2 = 0.2\,\text{kg}\,\text{m}^2$

EXAMPLE 19.2

Find the polar moment of inertia of a rim type flywheel whose dimensions
are as shown in Fig. 19.2. The density of the steel from which it is made is
7800 kg/m^3.

A rim type flywheel is a wheel in which
all of its mass may be considered to be
concentrated round the rim – the hub
and spokes being neglected.

All the mass may be considered to be
situated at one particular radius,
namely the mean radius of the rim.
This may be taken and used as the
radius of gyration k.

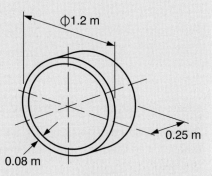

Fig. 19.2

Hence $\qquad$ Polar $k = 0.6 - 0.04 = 0.56$

$\qquad$ Volume of the rim $= 2\pi \times$ mean radius $\times$ width $\times$ thickness

$\qquad\qquad\qquad = 2\pi \times 0.56 \times 0.25 \times 0.08 = 0.0704\,\text{m}^3$

and $\qquad$ Mass of the rim $=$ volume $\times$ density

$\qquad\qquad\qquad = 0.0704 \times 7800 = 549\,\text{kg}$

Hence $\qquad$ Polar $I = $ total mass $\times k^2$

$\qquad\qquad\qquad = 549 \times 0.56^2 = 172\,\text{kg}\,\text{m}^2$

EXAMPLE 19.3

Find the moment of inertia, about its polar axis, of a solid cylinder of mass M and radius R.

The mass of a solid cylinder is not concentrated at a particular radius and so we shall consider that the cylinder is made up of a series of elementary rings (each similar to the rim of a rim type flywheel) as in Fig. 19.3.

Let the density of the material be ρ, and the length of the cylinder be l.

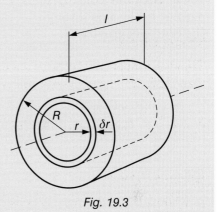

Fig. 19.3

Now for the cylinder

$$
\begin{aligned}
I &= \Sigma(mr^2) \\
&= \Sigma \text{ (mass of elementary ring} \times r^2) \\
&= \Sigma \text{ (volume of elementary ring} \times \text{density} \times r^2) \\
&= \Sigma \,(2\pi r\,\delta r\, l \times \rho \times r^2) \;=\; \Sigma(2\pi\rho l r^3\,\delta r) \\
&= \int_0^R 2\pi\rho l r^3\,\mathrm{d}r \;=\; 2\pi\rho l \int_0^R r^3\,\mathrm{d}r \;=\; 2\pi\rho l \left[\frac{r^4}{4}\right]_0^R \\
&= \frac{2\pi\rho l R^4}{4} \;=\; \tfrac{1}{2}(\pi R^2 l \rho)R^2 \\
&= \tfrac{1}{2}MR^2
\end{aligned}
$$

since the mass of the cylinder is $M = \pi R^2 l \rho$.

If we also wish to find the radius of gyration k about the polar axis we may use

$$I = Mk^2 \quad \text{since} \quad M = \Sigma m$$

$$\therefore \qquad \tfrac{1}{2}MR^2 = Mk^2$$

from which

$$k^2 = \frac{R^2}{2}$$

or

$$k = \frac{R}{\sqrt{2}}$$

PARALLEL AXIS THEOREM

This is similar to that for second moments of area.

In Fig. 19.4 the axis CD is parallel to the axis through the centre of gravity G.

The distance between the axes is h.

$$\boxed{I_{\text{CD}} = I_{\text{axis through G}} + Mh^2}$$

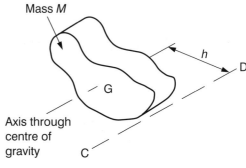

Mass M

h

D

G

Axis through centre of gravity

C

Fig. 19.4

EXAMPLE 19.4

Find the moment of inertia for the connecting rod shown in Fig. 19.5, about an axis through its centre of gravity G and parallel to the knife edge. Its mass is $41\,\text{kg}$ and I about the knife edge is $15.5\,\text{kg}\,\text{m}^2$ found by oscillating the connecting-rod as a compound pendulum.

Using the parallel axis theorem,

$$I_{\text{knife edge}} = I_{\text{axis through G}} + Mh^2$$

$$15.5 = I_{\text{axis through G}} + 41\left(\frac{530}{1000}\right)^2$$

$$15.5 = I_{\text{axis through G}} + 11.5$$

$$I_{\text{axis through G}} = 15.5 - 11.5$$

$$= 4\,\text{kg}\,\text{m}^2$$

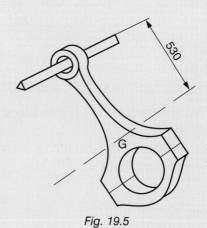

530

G

Fig. 19.5

EXPRESSIONS FOR MOMENTS OF INERTIA FOR COMMON SHAPES

These may be found in most engineering reference books, typical data being given as follows:

Solid Cylinder (Fig. 19.6)

Polar $\quad I_{OO} = \frac{1}{2}MR^2$

and $\quad I_{GG} = M\left(\dfrac{L^2}{12} + \dfrac{R^2}{4}\right)$

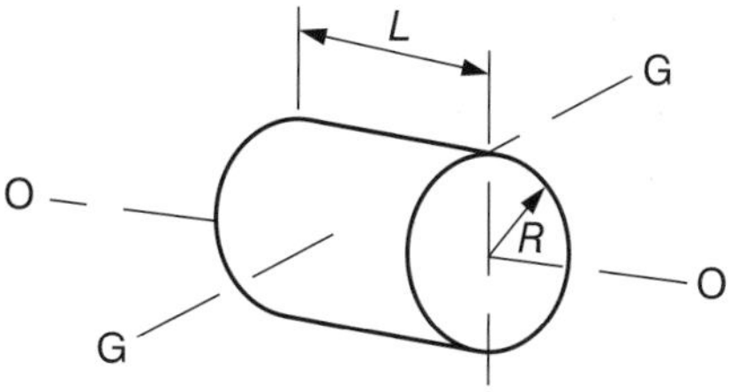

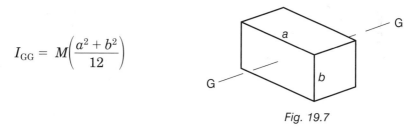

Fig. 19.6

where GG passes through the centre of gravity and is perpendicular to OO.

Solid Rectangular Block (Fig. 19.7)

$$I_{GG} = M\left(\dfrac{a^2 + b^2}{12}\right)$$

Fig. 19.7

where GG passes through the centre of gravity and is perpendicular to face ab.

Exercise 19.1

1) Find the polar moment of inertia of a solid aluminium cylinder 100 mm diameter and 500 mm long, if the density of aluminium is 2700 kg/m³.

2) Find the radius of gyration k_{GG} of the solid rectangular block shown in Fig. 19.8 if axis GG is perpendicular to the face measuring 200 mm × 300 mm and passes through the centre of gravity of the block.

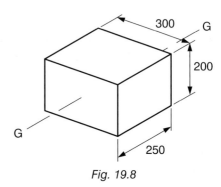

Fig. 19.8

3) Find the polar moment of inertia of a hollow steel cylinder 750 mm long, having an outside diameter of 500 mm and an inside diameter of 400 mm. The density of steel is 7900 kg/m³.

(*Hint*: The polar moment of inertia of the hollow cylinder is the difference between the polar moments of inertia of a solid cylinder 500 mm in diameter and a solid cylinder 400 mm in diameter.)

4) A hollow copper cylinder is as shown in Fig. 19.9. If the density of copper is 9000 kg/m³ find: **a)** I_{OO} **b)** I_{AA}

(*Hint*: When using the parallel axis theorem which gives $I_{AA} = I_{OO} + Mh^2$ remember that M is the mass of the hollow cylinder.)

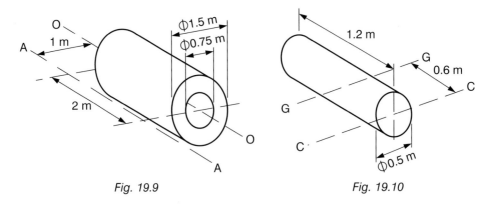

Fig. 19.9   *Fig. 19.10*

5) Figure 19.10 shows a solid copper cylinder which has a density of 9000 kg/m³. Find:

    a) I_{GG}, where GG is an axis passing through the centre of gravity and perpendicular to the polar axis

    b) I_{CC}

6) Find I_{BB} for the solid block shown in Fig. 19.11 if the density of the material from which it is made is 8000 kg/m³.

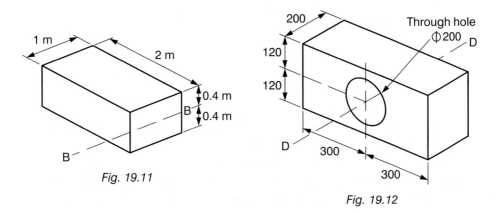

Fig. 19.11

Fig. 19.12

7) The component shown in Fig. 19.12 is made from a material which has a density of 7500 kg/m³. Find I_{DD}.

20 Numerical Integration

INTRODUCTION

We know that the area under a curve between limits $x = a$ and $x = b$ is given by the sum of all the elementary strip areas (Fig. 20.1).

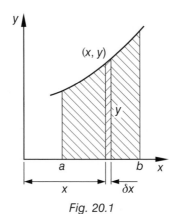

Fig. 20.1

In mathematical notation

$$\text{Area} = \sum_{x=a}^{x=b} y\,\delta x \qquad \text{approximately}$$

or

$$\text{Area} = \int_a^b y\,dx \qquad \text{exactly}$$

However, it is not always possible to evaluate the integral by direct mathematical integration. For example, we may have a curve obtained from experimental results for which there is no equation giving y in terms of x. Another difficulty arises when the expression for y in terms of x is too complicated for us to integrate.

We may then have to resort to dividing the required area into a number of strips (similar to elementary strips), finding the area of each strip, and then adding these up. Therefore the result will depend on calculating, as accurately as possible, the areas of the vertical strips.

Three reasonably simple methods are available: trapezium, mid-ordinate and Simpson's rules. You may have used these previously, but we shall now examine, in more detail, their derivation and the accuracy of results obtained.

The worked examples, which appear later in this chapter, to illustrate the use of the three methods can also be solved by direct integration and thus exact results obtained. (Here 'exact' means 'as accurate as required' since the answers are decimal numbers.) You may find it useful to check these. This enables the error to be calculated for each result.

THE TRAPEZIUM (OR TRAPEZOIDAL) RULE

Consider the area having the boundary ABCD shown in Fig. 20.2.

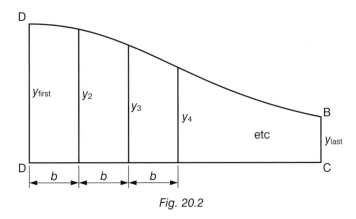

Fig. 20.2

The area is divided into a number of vertical strips of equal width b.

Each vertical strip is assumed to be a trapezium. Hence the third strip, for example, will have an area $= b \times \frac{1}{2}(y_3 + y_4)$.

But

$$\text{Area ABCD} = \text{The sum of all the vertical strips}$$

$$= b \times \tfrac{1}{2}(y_{\text{first}} + y_2) + b \times \tfrac{1}{2}(y_2 + y_3) + b \times \tfrac{1}{2}(y_3 + y_4) + \dots$$

$$= b[\tfrac{1}{2}y_{\text{first}} + \tfrac{1}{2}y_2 + \tfrac{1}{2}y_2 + \tfrac{1}{2}y_3 + \dots + \tfrac{1}{2}y_{\text{last}}]$$

$$= b[\tfrac{1}{2}(y_{\text{first}} + y_{\text{last}}) + y_2 + y_3 + y_4 \dots]$$

> Area ABCD = Strip width $\times \left[\tfrac{1}{2}(\text{Sum of first and last ordinates})\right.$
>
> $\left. + (\text{Sum of remaining ordinates})\right]$

This is known as the **trapezium rule**.

EXAMPLE 20.1

Find $\displaystyle\int_0^{1.5} e^x \, dx$ by numerical integration using the trapezium rule

The integral represents the area under the curve $y = e^x$ as shown in Fig. 20.3. The area has been divided into six vertical strips of equal width and the lengths of the ordinates found using a calculator.

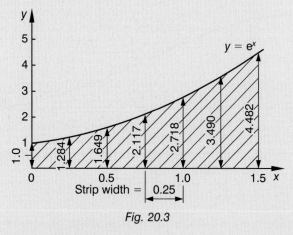

Fig. 20.3

Now the trapezium rules states

$$\text{Area} = (\text{Strip width}) \times \left[\tfrac{1}{2}(\text{Sum of first and last ordinates})\right.$$

$$\left. + (\text{Sum of remaining ordinates})\right]$$

Thus

$$\int_0^{1.5} e^x \, dx = \text{Area under the curve}$$

$$\approx 0.25 \times \left[\tfrac{1}{2}(1.000 + 4.482) + (1.284 + 1.649 + 2.117 + 2.718 \right.$$

$$\left. + 3.490) \right]$$

$$\approx 3.500 \text{ correct to 3 decimal places}$$

The result by integration is 3.482 correct to 3 decimal places.

$$\therefore \qquad \text{Error} = \frac{3.500 - 3.482}{3.482} \times 100 = +0.52\%$$

EXAMPLE 20.2

Find $\displaystyle\int_0^{\pi} \sin\theta \, d\theta$ by numerical integration using the trapezium rule

The integral represents the area under the curve $y = \sin\theta$ as shown in Fig. 20.4. The area has been divided into six vertical strips of equal width and the lengths of the ordinates found using a calculator.

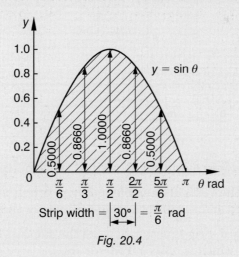

Fig. 20.4

Now the trapezium rule states

$$\text{Area} = (\text{Strip width}) \times \left[\tfrac{1}{2}(\text{Sum of first and last ordinates}) \right.$$

$$\left. + (\text{Sum of remaining ordinates}) \right]$$

Thus

$$\int_0^{\pi} \sin \theta \, d\theta = \text{Area under the curve}$$

$$\approx \frac{\pi}{6} \times \left[\tfrac{1}{2}(0 + 0) + (0.5000 + 0.8660 + 1.0000 + 0.8660 \right.$$

$$\left. + 0.5000) \right]$$

$$\approx 1.954 \text{ correct to 3 decimal places}$$

The result by integration is 2.000 correct to 3 decimal places.

$$\therefore \qquad \text{Error} = \frac{1.954 - 2.000}{2.000} \times 100 = -2.3\%$$

THE MID-ORDINATE RULE

Consider the area having the boundary ABCD shown in Fig. 20.5.

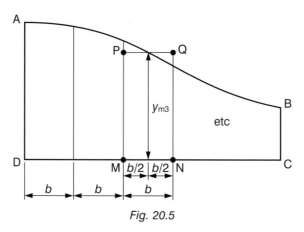

Fig. 20.5

The area is divided into a number of vertical strips of equal width b.

In the third strip the vertical line of length y_{m3} is half way between the boundary ordinates and is called the mid-ordinate of the third strip.

The area of the third strip is assumed to be the same as the area of the rectangle PQNM which is $(\text{MN} \times \text{PM})$ or by_{m3}.

Thus, if the mid-ordinates of all the strips are y_{m1}, y_{m2}, y_{m3}, etc.

then   Area ABCD $= by_{m1} + by_{m2} + by_{m3} + \ldots$

$$= b(y_{m1} + y_{m2} + y_{m3} + \ldots + y_{m \text{ last}})$$

or $\boxed{\text{Area ABCD} = (\text{Strip width}) \times (\text{Sum of mid-ordinates})}$

EXAMPLE 20.3

Find $\displaystyle\int_0^{1.5} e^x \, dx$ by numerical integration using the mid-ordinate rule

The integral represents the area under the curve $y = e^x$ as shown in Fig. 20.6. The area has been divided into six vertical strips of equal width and the lengths of the mid-ordinates found using a calculator.

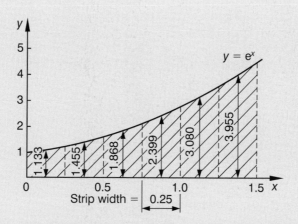

Fig. 20.6

Now the mid-ordinate rule states

$$\text{Area} = (\text{Strip width}) \times (\text{Sum of mid-ordinates})$$

Thus

$$\int_0^{1.5} e^x \, dx = \text{Area under the curve}$$

$$\approx 0.25 \times (1.133 + 1.455 + 1.868 + 2.399 + 3.080 + 3.955)$$

$$\approx 3.473 \text{ correct to 3 decimal places}$$

The result by integration is 3.482 correct to 3 decimal places.

$$\therefore \quad \text{Error} = \frac{3.473 - 3.482}{3.482} \times 100 = -0.26\%$$

EXAMPLE 20.4

Find $\int_0^\pi \sin\theta\,d\theta$ by numerical integration using the mid-ordinate rule

The integral represents the area under the curve $y = \sin\theta$ as shown in Fig. 20.7. The area has been divided into six vertical strips of equal width and the lengths of the mid-ordinates found using a calculator.

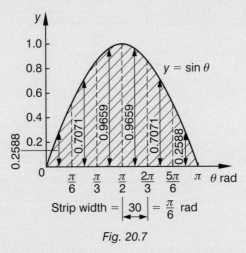

Fig. 20.7

Now the mid-ordinate rule states

   Area = (Strip width) × (Sum of mid-ordinates)

Thus

$\int_0^\pi \sin\theta\,d\theta$ = Area under the curve

$\approx \dfrac{\pi}{6} \times (0.2588 + 0.7071 + 0.9659 + 0.9659 + 0.7071 + 0.2588)$

≈ 2.023 correct to 3 decimal places

The result by integration is 2.000 correct to 3 decimal places.

∴   Error = $\dfrac{2.023 - 2.000}{2.000} \times 100 = +1.15\%$

SIMPSON'S RULE

Suppose we wish to find the area PQRNL shown shaded in Fig. 20.8. QM is the mid-ordinate which divides the area into two strips of equal width b. Figure 20.9 is a repeat of Fig. 20.8 but shows a straight line AB drawn through point Q and parallel to chord PR.

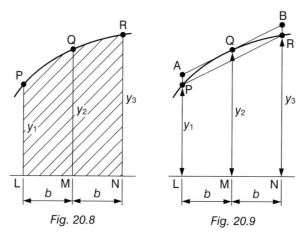

Fig. 20.8   Fig. 20.9

Now

$$\begin{pmatrix} \text{Area of} \\ \text{parallelogram ABRP} \end{pmatrix} = \begin{pmatrix} \text{Area of} \\ \text{trapezium ABNL} \end{pmatrix} - \begin{pmatrix} \text{Area of} \\ \text{trapezium PRNL} \end{pmatrix}$$

$$= 2by_2 - \tfrac{1}{2}(y_1 + y_3)2b$$

$$= b(2y_2 - y_1 - y_3)$$

Now the area PQR, between the curve and the chord, is taken as $\tfrac{2}{3}$ of the area of the parallelogram ABRP. This is a very close approximation. (It is only exact if the curve is the arc of a parabola and also AB is the tangent at Q.)

Thus   Area PQRNL $= \begin{pmatrix} \text{Area of} \\ \text{trapezium PRNL} \end{pmatrix} + \text{Area PQR}$

$$= \tfrac{1}{2}(y_1 + y_3)2b + \tfrac{2}{3}b(2y_2 - y_1 - y_3)$$

$$= b(y_1 + y_3 + \tfrac{4}{3}y_2 - \tfrac{2}{3}y_1 - \tfrac{2}{3}y_3)$$

$$= \frac{b}{3}(y_1 + 4y_2 + y_3)$$

Figure 20.10 shows a larger area divided into six vertical strip areas of equal width. Each adjacent pair of strips forms an area similar to that shown in Fig. 20.8. We therefore apply the expression we derived for each pair of strips as follows:

$$\text{Shaded area} = \begin{pmatrix} \text{Area between} \\ \text{ordinates} \\ y_1 \text{ and } y_3 \end{pmatrix} + \begin{pmatrix} \text{Area between} \\ \text{ordinates} \\ y_3 \text{ and } y_5 \end{pmatrix} + \begin{pmatrix} \text{Area between} \\ \text{ordinates} \\ y_5 \text{ and } y_7 \end{pmatrix}$$

$$= \frac{b}{3}(y_1 + 4y_2 + y_3) + \frac{b}{3}(y_3 + 4y_4 + y_5) + \frac{b}{3}(y_5 + 4y_6 + y_7)$$

$$= \frac{b}{3}\left[y_1 + y_7 + 4(y_2 + y_4 + y_6) + 2(y_3 + y_5)\right]$$

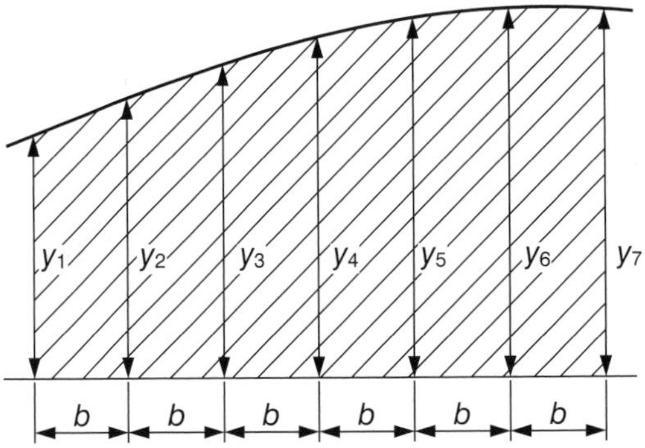

Fig. 20.10

This idea may be extended providing there is an **even number of strips** of equal width to give a general expression for Simpson's rule:

$$\text{Area} = \tfrac{1}{3}(\text{Strip width}) \times \Big[(\text{Sum of first and last ordinates})$$
$$+ 4(\text{Sum of even ordinates})$$
$$+ 2(\text{Sum of remaining odd ordinates})\Big]$$

EXAMPLE 20.5

Find $\int_0^{1.5} e^x\,dx$ by numerical integration using Simpson's rule

In order to use Simpson's rule the required area must be divided into an *even* number of strips. This has been done in Fig. 20.3 and we will use the values of the ordinates given.
Now Simpson's rule states

$$\text{Area} = \tfrac{1}{3}(\text{Strip width}) \times \Big[(\text{Sum of first and last ordinates})$$

$$+ 4(\text{Sum of even ordinates})$$

$$+ 2(\text{Sum of remaining odd ordinates})\Big]$$

Thus

$$\int_0^{1.5} e^x\,dx = \text{Area under the curve}$$

$$\approx \tfrac{1}{3} \times 0.25 \times \Big[1.000 + 4.482)$$

$$+ 4(1.284 + 2.117 + 3.490)$$

$$+ 2(1.649 + 2.718)\Big]$$

$$\approx 3.482 \text{ correct to 3 decimal places}$$

The result by integration is 3.482 correct to 3 decimal places and so there is no error.

An alternative layout using a table to show the calculations is often used.

Ordinate number	Length	Simpson's multiplier	Product
1	1.000	1	1.000
2	1.284	4	5.136
3	1.649	2	3.298
4	2.117	4	8.468
5	2.718	2	5.436
6	3.490	4	13.960
7	4.482	1	4.482
		Total product = 41.780	

Hence   Result $= \tfrac{1}{3} \times 0.25 \times 41.780 = 3.482$

EXAMPLE 20.6

Find $\displaystyle\int_0^\pi \sin\theta\,d\theta$ by numerical integration using Simpson's rule

In order to use Simpson's rule the required area must be divided into an *even* number of strips. This has been done in Fig. 20.4 and we will use the values of the ordinates given.

Now Simpson's rule states

$$\text{Area} = \tfrac{1}{3}(\text{Strip width}) \times \Big[(\text{Sum of first and last ordinates})$$

$$+ 4(\text{Sum of even ordinates})$$

$$+ 2(\text{Sum of remaining odd ordinates})\Big]$$

Thus

$$\int_0^\pi \sin\theta\,d\theta = \text{Area under the curve}$$

$$\approx \frac{1}{3}\times\frac{\pi}{6}\times\Big[(0+0) + 4(0.5000 + 1.0000 + 0.5000)$$

$$+ 2(0.8660 + 0.8660)\Big]$$

$$\approx 2.001 \ \text{correct to 3 decimal places}$$

The result by integration is 2.000 correct to 3 decimal places.

$$\therefore \quad \text{Error} = \frac{2.001 - 2.000}{2.000}\times 100 = 0.05\%$$

COMPARISON OF TRAPEZIUM, MID-ORDINATE AND SIMPSON'S RULES

If a curve is shaped ⌒ then the trapezium rule gives the shaded area shown in Fig. 20.11. This area is smaller than the true area under the curve, and this is conformed by the answer to Example 20.2 which is 2.3% too small.

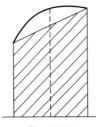

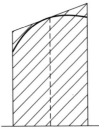

Fig. 20.11   Fig. 20.12

The mid-ordinate rule gives the shaded area shown in Fig. 20.12, which is greater than the true area under the curve. This is confirmed by the answer to Example 20.4 which is 1.15% too large.

For a curve shaped ⌣ the trapezium rule gives results too large (see Example 20.1), whilst the mid-ordinate rule gives results too small (see Example 20.3).

Simpson's rule, which is a combination of both the other rules, is much more accurate, as confirmed by the results of Examples 20.5 and 20.6. Hence if accuracy is of importance then Simpson's rule should be used.

ACCURACY OF ANSWERS USING SIMPSON'S RULE

In the preceding text it was possible to obtain exact answers to the worked examples. This enabled us to compare results obtained using approximate methods so that we may gain some idea of the degree of accuracy which may be expected.

However, in practice, we would only use Simpson's rule for integration if it were *not* possible to get an exact answer.

The accuracy of a result using Simpson's rule will depend on the number of intervals chosen. There are expressions which may be used to estimate the error, and also to decide on the number of intervals required to obtain an answer to a particular degree of accuracy. However, these are tedious to use and it is recommended that the following method be used:

> If an increase in the number of equal intervals does not involve a change in the answer to a certain degree of accuracy, then we may rely on the result to that degree of accuracy.

Suppose, for example, that for a particular problem when using Simpson's rule we obtain (correct to 3 significant figures):

with 6 equal intervals an answer of 6.36
with 8 equal intervals an answer of 6.34
with 10 equal intervals an answer of 6.33
and   with 12 equal intervals an answer of 6.33

Here we may safely assume the result is 6.33 correct to 3 significant figures.

Exercise 15.1

Evaluate, stating the answers to 3 significant figures, the following integrals, using:

a) the trapezium rule
b) the mid-ordinate rule
c) Simpson's rule.

1) $\displaystyle\int_0^3 \sqrt{(9 - x^2)}\,dx$ using 6 intervals

2) $\displaystyle\int_0^{1.5} \frac{dx}{\sqrt{(4 - x^2)}}$ using 6 intervals

3) $\displaystyle\int_1^2 (\log_e x)\,dx$ using 10 intervals

4) $\displaystyle\int_0^{\pi/4} \left(\frac{\theta}{\cos\theta}\right) d\theta$ using 6 intervals

5) $\displaystyle\int_{0.1}^{0.7} \left(\frac{t}{1 - t}\right) dt$ using 6 intervals

6) $\displaystyle\int_0^{\pi/6} (\sin^3 \phi)\,d\phi$ using 6 intervals

7) $\displaystyle\int_0^2 (x^2 e^{-x})\,dx$ using 8 intervals

8) Use Simpson's rule to find the value of

$$\int_{1.8}^3 \left(\frac{1}{\sin x}\right) dx$$

giving the answer correct to 3 significant figures. Start with 6 equal intervals and proceed until the required accuracy is guaranteed.

Probability 21

Simple, empirical and total probability – the probability scale – mutually exclusive events – the addition law – independent events – the multiplication law – the binomial distribution – the poisson distribution.

INTRODUCTION

Probability is the chance of something happening. One obvious area where this is of extreme importance is in the insurance industry. For instance the greater the chance that your house will be burgled then the higher will be the premium you pay.

It may not seem quite so obvious but probability is also of interest in technology. In this chapter we shall generally be referring to the estimating of events happening in engineering production.

This is implicit in production quality control and we shall learn about some useful techniques and their application.

SIMPLE PROBABILITY

If a coin is tossed it will come down either heads or tails. There are only these two possibilities. The probability of obtaining a head in a single toss is one possibility out of two possibilities. We write this:

$$\Pr(\text{head}) = \frac{\text{possibility of a head}}{\text{total possibilities}} = \frac{1}{2}$$

When we roll a die (plural: dice) we can get one of six possible scores, 1, 2, 3, 4, 5 or 6. The probability of scoring 3 in a single roll of the die is one possibility out of a total of six possibilities. Hence:

$$\Pr(\text{three}) = \frac{\text{possibility of a three}}{\text{total possibilities}} = \frac{1}{6}$$

There are 52 playing cards in a pack. When we cut the pack we can obtain one of these cards and hence the total possibilities are 52. Since there are four aces in the pack:

$$\Pr(\text{ace}) = \frac{\text{possibility of an ace}}{\text{total possibilities}} = \frac{4}{52} = \frac{1}{13}$$

EXAMPLE 21.1

Calculate the probability of cutting a king, queen or jack in a single cut of a pack of cards.

$$\text{Total possibilities} = 52$$

$$\text{Possibility of cutting a king, queen or jack} = 12$$

$$\text{Pr(king, queen or jack)} = \frac{12}{52} = \frac{3}{13}$$

THE PROBABILITY SCALE

When an event is absolutely certain to happen we say that the probability of it happening is 1. Thus the probability that one day each of us will die is 1. When an event can never happen we say that the probability of it happening is 0. Thus the probability that any one of us can jump 5 metres high unaided is 0. All probabilities must, therefore, have a value between 0 and 1. They can be expressed as either a fraction or a decimal. Thus:

$$\text{Pr(head)} = \frac{1}{2} = 0.5 \qquad \text{Pr(ace)} = \frac{1}{13} = 0.077$$

Any probability can be calculated from the formula:

$$p = \frac{\text{number of ways in which a event can happen}}{\text{total possibilities}}$$

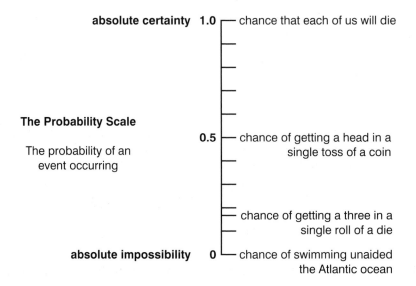

EXAMPLE 21.2

In two successive throws of a die what is the probability that the sum of the scores will be eight?

If we examine the possible results
we see that the following
results can occur:

1, 1	2, 1	3, 1	4, 1	5, 1	6, 1
1, 2	2, 2	3, 2	4, 2	5, 2	6, 2
1, 3	2, 3	3, 3	4, 3	5, 3	6, 3
1, 4	2, 4	3, 4	4, 4	5, 4	6, 4
1, 5	2, 5	3, 5	4, 5	5, 5	6, 5
1, 6	2, 6	3, 6	4, 6	5, 6	6, 6

We see that the total possibilities equals 36. The following results total eight: 2, 6; 3, 5; 4, 4; 5, 3; 6, 2. Hence the number of ways in which the event of scoring eight can happen is 5. Hence:

$$\Pr(\text{eight}) \ = \ \frac{5}{36}$$

EMPIRICAL PROBABILITY

When rolling a die we expect each face to turn up in one-sixth of the number of throws. In any particular series of throws we would be rather surprised if we obtained exactly this result. However by making a large number of throws we shall get very near to one-sixth. This suggests another way of calculating the probability of an event happening:

$$p \ = \ \frac{\text{total number of the occurrences of the event}}{\text{total number of trials}}$$

The probability calculated by this formula is an empirical probability because it arises as a result of a statistical experiment.

For very many events it is not possible to obtain a theoretical probability and in such cases we resort to an empirical probability.

EXAMPLE 21.3

2000 parts were checked by measuring them and it was found that 100 were outside of the drawing limits (i.e. they were defective). What is the probability of finding a defective part in a single trial?

$$\Pr(\text{defective part}) \ = \ \frac{\text{total number of defective parts found}}{\text{total number checked}}$$

$$= \ \frac{100}{2000} = 0.05$$

Probabilities are often expressed as percentages and in this case we would say:

$$\text{percentage defective} \ = \ 0.05 \times 100 \ = \ 5\%$$

TOTAL PROBABILITY

If p = the probability of an event happening

and q = the probability of it not happening

then $p + q = 1.$

That is,

the total probability, covering all possible events is always equal to 1.

Thus in tossing a coin if we regard a head as a success then $p = \frac{1}{2}$. The event of obtaining a tail is then regarded as a failure and we say $q = \frac{1}{2}$. Thus the total probability, i.e. the probability of obtaining a head or a tail, is $\frac{1}{2} + \frac{1}{2} = 1$.

Again, if in rolling a die we regard the rolling of a three as a success then $p = \frac{1}{6}$. Any other face of the die turning up will be regarded as a failure and hence $q = \frac{5}{6}$.

The total probability covering all possible events is $\frac{1}{6} + \frac{5}{6} = 1$

EXAMPLE 21.4

A bag contains 4 red balls, 3 blue, 2 green and one yellow. Find the probability of drawing a red ball in a single draw from the bag. What is the probability of not drawing a red ball?

$$\text{Pr(red)} = \frac{\text{number of ways of drawing a red ball}}{\text{total number of possibilities}} = \frac{4}{10} = 0.4$$

Since the probability of drawing a red ball is 0.4,

$$\text{Pr(not red)} = 1 - 0.4 = 0.6$$

Exercise 21.1

Find the probability of each of the following events occurring:

1) An ace, king, queen or jack of diamonds appearing when a pack of 52 playing cards is cut.

2) The sum of nine appearing when two dice are thrown simultaneously.

3) A silver coin being chosen at random from a purse containing 8 copper coins and 5 silver coins.

4) A bag contains 8 red, 5 white, 4 black and 3 green balls. A ball is drawn at random from the bag and replaced. Calculate the probability that it will be **a)** red, **b)** white, **c)** not red, **d)** not green, **e)** black.

5) What is the probability that an even number will appear in a single roll of a die?

6) 1000 parts were checked by means of a limit gauge and 20 were found to fail the test. What is **a)** the percentage defective, **b)** the probability of finding a defective part in a single trial?

7) 20 ordinary bolts became accidently mixed with 180 high strength bolts. Calculate the probability of choosing a high strength bolt when one bolt is chosen at random from the 200 bolts.

ADDITION LAW OF PROBABILITY

If two or more events are such that not more than one of them can occur in a single trial they are said to be *mutually exclusive*.

Thus the events of throwing a 5 or a 6 in a single roll of a die are mutually exclusive events because it is only possible to throw either a 5 or a 6; it is impossible to throw both.

If p_1, p_2, p_3, ... are the separate probabilities of the occurrence of 1, 2, 3, ... mutually exclusive events then the probability that *one* of these events will occur is:

$$P = p_1 + p_2 + p_3 + ...$$

EXAMPLE 21.5

Find the probability of drawing either an ace, or the king of diamonds, or the queen of hearts, in a single cut of a pack of 52 playing cards.
In a single trial,

$$\text{Pr(ace)} = p_1 = \frac{4}{52} = \frac{1}{13}$$

$$\text{Pr(king of diamonds)} = p_2 = \frac{1}{52}$$

$$\text{Pr(queen of hearts)} = p_3 = \frac{1}{52}$$

$$\text{Pr(ace, king of diamonds or queen of hearts)} = p_1 + p_2 + p_3$$

$$= \frac{1}{13} + \frac{1}{52} + \frac{1}{52}$$

$$= \frac{6}{52} = 0.12$$

Note that the events of drawing an ace, the king of diamonds and the queen of hearts in a single cut are mutually exclusive events because only one of the three events can occur.

The ***addition law*** is sometimes called the ***or law*** because it is the probability of the occurrence of one event ***or*** another event that is required. In Example 21.5 it was the probability of drawing an ace ***or*** the king of diamonds ***or*** the queen of hearts that was required. Notice that when the addition law is used the chances of one of the events happening is ***increased***.

EXAMPLE 21.6

It is known that the probability of obtaining 0 defectives in a sample of 40 items is 0.34, whilst the probability of obtaining 1 defective part is 0.46. What is the probability of obtaining:
a) not more than 1 defective part in a sample,
b) more than 1 defective part in a sample?

If we choose a random sample of 40 items then this sample may contain any number of defective parts in it up to a maximum of 40. The events of drawing samples with certain numbers of defectives in them are mutually exclusive events.

a) Pr(not more than 1 defective) = the probability of drawing a sample with 0 defective parts in it + the probability of drawing a sample with 1 defective part in it

$$= 0.34 + 0.46 = 0.80$$

b) Since the total probability covering all possible events is 1,
Pr(more than 1 defective) $= 1 - 0.80 = 0.20$

MULTIPLICATION LAW

Two or more events are said to be independent if the probability of the occurrence of any one of the events is not influenced by the occurrence of any other of the events.

In two separate cuts of a pack of cards, what happens on the first cut in no way affects what happens on the second cut. Hence these are independent events.

Similarly in two horse races, what happens in the first race in no way affects what happens in the second race and hence the two events are independent.

If p_1, p_2, p_3, ... are the separate probabilities of the occurrence of 1, 2, 3, ... independent events the probability that *all* of the events will happen is:

$$P = p_1 \times p_2 \times p_3 \ldots$$

EXAMPLE 21.7

If three dice are thrown simultaneously, find the probability that each will show six.

The events of throwing the three dice are independent events and

on the first die, $\text{Pr(six)} = p_1 = \dfrac{1}{6}$

on the second die, $\text{Pr(six)} = p_2 = \dfrac{1}{6}$

on the third die, $\text{Pr(six)} = p_3 = \dfrac{1}{6}$

$$\text{Pr(three sixes)} = p_1 \times p_2 \times p_3 = \frac{1}{6} \times \frac{1}{6} \times \frac{1}{6} = \frac{1}{216} = 0.0046$$

The **_multiplication law_** is sometimes called the **_and_** law because it is the probability of one event **_and_** another event which is required.
In Example 21.7 it was the probability of throwing a six on the first die **_and_** throwing a six on the second die **_and_** throwing a six on the third die that was required. Notice that when the multiplication law is used the chances that all the events will happen are **_reduced_**.

EXAMPLE 21.8

It is known that 10% of the items produced on a certain machine tool are defective. If a sample of three items is chosen at random from a large batch of these items, what is the probability that all three items will be defective?

Since what happens on the first draw in no way affects what happens on the other draws, the three events are independent.

On the first draw, $\text{Pr(defective)} = p_1 = 0.1$

On the second draw, $\text{Pr(defective)} = p_2 = 0.1$

On the third draw, $\text{Pr(defective)} = p_3 = 0.1$

     $\text{Pr(three defectives)} = p_1 \times p_2 \times p_3 = 0.1 \times 0.1 \times 0.1 = 0.001$

It is possible to have a mixture of independent and dependent events. For instance, suppose in Example 21.8 we desire to know the probability of drawing only one defective part in a sample of three items.

There are three distinct possibilities as shown in the table below:

	1st draw	2nd draw	3rd draw
1st possibility	D	G	G
2nd possibility	G	D	G
3rd possibility	G	G	D

Any of these three possibilities gives only one defective in the sample. Each of the three possibilities are mutually exclusive events and hence:

$$\text{Pr(one defective)} = (0.1 \times 0.9 \times 0.9) + (0.9 \times 0.1 \times 0.9) + (0.9 \times 0.9 \times 0.1)$$
$$= 3 \times 0.1 \times (0.9)^2 = 0.243$$

EXERCISE 21.2

1) Find the probability of drawing either an ace, king or queen in a single draw from a pack of 52 playing cards.

2) The probability of obtaining 0 defective items in a sample is 0.20 whilst the probability of obtaining 1 defective part in the same size of sample is 0.25. Calculate the probability of drawing a sample with not more than one defective part in it.

3) A ball is drawn from a bag containing 3 red, 5 blue and 2 white balls. What is the probability that it will be either red or white?

4) Calculate the probability of rolling a 3 or a 6 in a single roll of a die.

5) If two fair dice are thrown at the same time, what is the probability of obtaining a score of 12?

6) Three ten pence coins are tossed together.

   Copy and complete the table to show the ways in which the coins could land.

Coin 1	Coin 2	Coin 3
H	H	H

   a) From your table write down the probability of:
      (i) the three coins all falling tails,
      (ii) two coins falling heads and one coin tails.
   b) The third coin is now changed for a six sided die and the two remaining coins and the die are tossed together. Calculate the probability that the throw will give:
      (i) tail, tail, six   (ii) one head, one tail and either a 1 or a 2.

7) A pound coin, a 50 p coin and a 10 p piece are tossed 200 times. For any particular toss find the probabilities that, assuming the coins are fair:
   a) both copper coins are heads,   b) all three coins are heads,
   c) at least one coin is heads,   d) at least two coins are heads.

8) A false coin is loaded so that on any one toss the probability of a head is $\frac{2}{3}$. The coin is tossed three times. Calculate the probabilities of obtaining:
   a) three heads,   b) three tails,
   c) two tails,   d) two heads.

9) A and B are points in a toy train system (Fig. 21.1). The probability of going straight on at each point is $\frac{2}{3}$. Find the probability that:

a) the train T hits the waiting train

b) the train T goes into the shed.

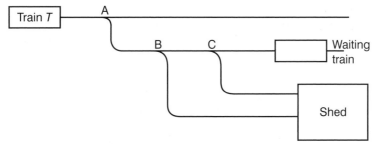

Fig. 21.1

10) The pointer shown in Fig. 21.2 is spun. Find the probability that the pointer:

a) stops in section B,

b) stops in either R or G section.

The result of two successive spins are noted. Find the probability that:

c) the first spin is an R, and the second spin is a G,

d) the pointer stops in either
 R then G   or   G then R.

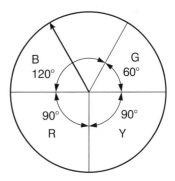

Fig. 21.2

11) A loaded die shows scores with the following probabilities:

Score	1	2	3	4	5	6
Probability	0.14	0.18	0.16	0.15	0.17	0.20

a) On a single throw what is the probability of a score less than 3?

b) If the die is thrown twice what is the probability of a 4 followed by a 6?

c) The die is thrown twice and the two scores added together. What is the probability that the total score will be:
   (i) 2   (ii) 12   (iii) exactly 7   (iv) less than 5?

12) A bag contains 14 yellow counters and 10 red counters. Two counters are taken out in succession and not replaced. Calculate the probability that:

a) both will be red,

b) both will be yellow,

c) the first will be red and the second yellow,

d) the first will be yellow and the second red.

REPEATED TRIALS

Suppose that it is known that 10% of components produced on a certain machine are defective. If we choose one component at random from a large batch of these components then the probability of it being defective is $p = 10/100 = 0.1$

The probability of it being good is $q = 90/100 = 0.9$

Since the component will have to be classified as either good or defective we have $p + q = 1$

Now from the same large batch let us choose two components at random. There are three distinct possibilities that can occur:

(a) Both the components may be good. The probability of this occurring is

$$(q) \times (q) \ = \ q^2 \ = \ (0.9)^2 \ = \ 0.81$$

by using the multiplication law. Another way of saying this is to state that the probability of obtaining 0 defectives is 0.81

(b) One component is good whilst the other is defective. This can occur in two ways: either the first component chosen is good with the second component defective, or the first component will be defective with the second good. By the multiplication law the probability of the first way occurring will be $q \times p = qp$, whilst the probability of the second way occurring will be $p \times q = pq$. If we are concerned only with the final result irrespective of the order in which the defectives occur then the probability of obtaining 1 defective is:

$$qp + pq \ = \ 2pq \ = \ 2 \times 0.1 \times 0.9 \ = \ 0.18$$

by using the addition law.

(c) Both components are defective. The probability of this occurring is:

$$(p) \times (p) = p^2 = (0.1)^2 = 0.01$$

Tabulating these results:

Possibility	Ways of arising	Probability of way	Probability of possibility occurring
0 defectives	good–good	q^2	q^2
1 defective	good–bad bad–good	qp pq	$2qp$
2 defectives	bad–bad	p^2	p^2

The terms in the table will be recognised as those arising from the expansion of $(q + p)^2$. Now let us analyse the results which will be obtained when taking a sample of 3 components.

The terms in the last column of the table will be recognised as those arising from the expansion of $(q + p)^3$.

Possibility	Ways of arising	Probability of way	Probability of possibility occurring
0 defectives	good–good–good	qqq	q^3
1 defective	good–good–defective good–defective–good defective–good-good	qqp qpq pqq	$3q^2p$
2 defectives	good–defective–defective defective–good–defective defective–defective–good	qpp pqp ppq	$3qp^2$
3 defectives	defective–defective–defective	ppp	p^3

From the foregoing we will rightly expect that the probabilities of obtaining 0, 1, 2, 3 and 4 defectives in a sample of 4 components would be the successive terms of the expansion of $(q + p)^4$.

Now, $(q + p)^4 = q^4 + 4q^3p + 6q^2p^2 + 4qp^3 + p^4$

Thus, if $p = 0.1$ and $q = 0.9$:

the probability of 0 defectives $= q^4 = (0.9)^4 = 0.6561$

the probability of 1 defective $= 4q^3p = 4 \times (0.9)^3 \times (0.1) = 0.2916$

the probability of 2 defectives $= 6q^2p^2 = 6 \times (0.9)^2 \times (0.1)^2 = 0.0486$

the probability of 3 defectives $= 4qp^3 = 4 \times (0.9) \times (0.1)^3 = 0.0036$

the probability of 4 defectives $= p^4 = (0.1)^4 = \underline{0.0001}$

the total probability covering all the possible results $= 1.0000$

We can now deduce the general rule.

The probabilities of obtaining 0, 1, 2, 3, ... n defectives in a sample of n items is given by the successive terms of the expansion of $(q + p)^n$.

In using this general rule it must be clearly understood that:

$n =$ number in the sample,

$p =$ probability of finding a defective in a single trial,

$q =$ probability of finding a good item in a single trial,

and **$p + q = 1$**

The expansion of $(q + p)^n$ can be obtained either by using the binomial theorem (see Chapter 5) or the table of binomial coefficients.

BINOMIAL COEFFICIENTS

The following table gives the binomial coefficients. The values are as found using Pascal's triangle but arranged differently.

The extreme left hand column is the value of n in $(q + p)^n$, and the top row is the number r of the particular term in the expression.

r:	0	1	2	3	4	5	6	7	8	9	10
$n =$											
1	1	1									
2	1	2	1								
3	1	3	3	1							
4	1	4	6	4	1						
5	1	5	10	10	5	1					
6	1	6	15	20	15	6	1				
7	1	7	21	35	35	21	7	1			
8	1	8	28	56	70	56	28	8	1		
9	1	9	36	84	126	126	84	36	9	1	
10	1	10	45	120	210	252	210	120	45	10	1

Note that the coefficients are symmetrical. Thus for example

$$(a + b)^5 = a^5 + 5a^4b + 10a^3b^2 + 10a^2b^3 + 5ab^4 + b^5$$

EXAMPLE 21.9

It is known that 10% of certain articles manufactured are defective. Find the probabilities of obtaining 0, 1, 2, 3, 4 and 5 defectives in a sample of 5 articles taken at random from a large batch of these articles.

The probability of a single article, chosen at random, being defective is $p = 0.1$. The probability of it being satisfactory is $q = 1 - 0.1 = 0.9$.

Now: $(q + p)^5 = q^5 + 5q^4p + 10q^3p^2 + 10q^2p^3 + 5qp^4 + p^5$

Number of defectives	Term of expansion	Probability	
0	q^5	$(0.9)^5$	$= 0.590\,49$
1	$5q^4p$	$5 \times (0.9)^4 \times (0.1)$	$= 0.328\,05$
2	$10q^3p^2$	$10 \times (0.9)^3 \times (0.1)^2$	$= 0.072\,90$
3	$10q^2p^3$	$10 \times (0.9)^2 \times (0.1)^3$	$= 0.008\,10$
4	$5qp^4$	$5 \times (0.9) \times (0.1)^4$	$= 0.000\,45$
5	p^5	$(0.1)^5$	$= 0.000\,01$

Total probability covering all possible results $= 1.000\,00$

The results are shown in Fig. 21.3 as a histogram.

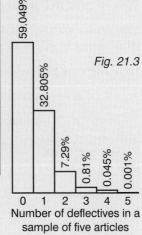

Fig. 21.3

59.049%
32.805%
7.29%
0.81%
0.045%
0.001%

0 1 2 3 4 5
Number of deflectives in a sample of five articles

EXAMPLE 21.10

A process is known to produce 5% of faulty articles. Estimate the chance of a sample of 12 articles **a)** containing exactly 3 faulty articles, **b)** containing more than 3 faulty articles.

Here $p = 0.05$ and $q = 0.95$ We need the first 4 terms of the expansion of $(q + p)^{12}$.

Number of defectives	Term of expansion	Probability	
0	q^{12}	$(0.95)^{12}$	$= 0.5400$
1	$12q^{11}p$	$12 \times (0.95)^{11} \times (0.05)$	$= 0.3411$
2	$66q^{10}p^2$	$66 \times (0.95)^{10} \times (0.05)^2$	$= 0.0987$
3	$220q^9p^3$	$220 \times (0.95)^9 \times (0.05)^3$	$= 0.0173$
		Total probabilities of above cases	$= 0.9971$

a) The probability of obtaining exactly 3 defectives is 0.0173 or 1.73%.
b) Since the total probability is 1, the probability of obtaining more than 3 defectives is $1 - 0.9971 = 0.0029$ or 0.29%.

THE BINOMIAL DISTRIBUTION

When a frequency distribution is obtained by use of the expansion of $(q + p)^n$ it is called a binomial distribution. By using the successive terms of $(q + p)^n$ a theoretical frequency table can be drawn up. A histogram can then be drawn which will show the distribution.

EXAMPLE 21.11

Draw a histogram for the expected frequencies obtained in tossing 10 coins 1024 times.

Here $p = q = \frac{1}{2}$ and $1024 = 2^{10}$:

$$(q + p)^{10} = q^{10} + 10q^9p + 45q^8p^2 + 120q^7p^3 + 210q^6p^4 + 252q^5p^5$$
$$+ 210q^4p^6 + 120q^3p^7 + 45q^2p^8 + 10qp^9 + p^{10}$$

The histogram for the frequency table is shown in Fig. 21.4 and it is seen to be symmetrical. A histogram for a binomial distribution is symmetrical if, and only if, $p = q = \frac{1}{2}$. In Fig. 21.3 the histogram is skewed to one side, the values of p and q being 0.1 and 0.9 respectively.

Number of Heads	Term from expansion	Frequency
0	$(\frac{1}{2})^{10} = \dfrac{1}{1024}$	1
1	$10 \times (\frac{1}{2})^9(\frac{1}{2}) = \dfrac{10}{1024}$	10
2	$45 \times (\frac{1}{2})^8(\frac{1}{2})^2 = \dfrac{45}{1024}$	45
3	$120 \times (\frac{1}{2})^7(\frac{1}{2})^3 = \dfrac{120}{1024}$	120
4	$210 \times (\frac{1}{2})^6(\frac{1}{2})^4 = \dfrac{210}{1024}$	210
5	$252 \times (\frac{1}{2})^5(\frac{1}{2})^5 = \dfrac{252}{1024}$	252
6	$210 \times (\frac{1}{2})^4(\frac{1}{2})^6 = \dfrac{210}{1024}$	210
7	$120 \times (\frac{1}{2})^3(\frac{1}{2})^7 = \dfrac{120}{1024}$	120
8	$45 \times (\frac{1}{2})^2(\frac{1}{2})^8 = \dfrac{45}{1024}$	45
9	$10 \times (\frac{1}{2})(\frac{1}{2})^9 = \dfrac{10}{1024}$	10
10	$(\frac{1}{2})^{10} = \dfrac{1}{1024}$	1

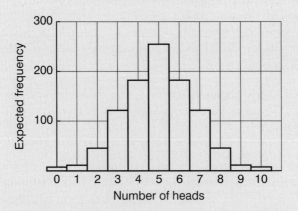

It can be shown that if np is less than 5 the histogram showing the expected frequency distribution will be skewed. If np is greater than 5 the histogram is reasonably symmetrical.

Exercise 21.3

1) It is known that in a large quantity of articles 10% are defective. Find the probabilities of obtaining 0, 1, 2, 3 or 4 defectives in a sample of 4 articles.

2) A machine is known to produce 5% of faulty articles. Find the probability of sample of 20 articles containing: **a)** no faulty articles, **b)** more than one faulty article.

3) A class of 20 students is prepared for an examination. Experience shows that 30% of the students fail. Find the probability of exactly 6 failing the examination.

4) A product is being made in large quantities. Successive samples of 40 items give the following numbers of defectives: 2, 2, 0, 0, 1, 3, 0, 2, 0, 0. Find the probability of obtaining in a sample of 40 items: **a)** no defectives, **b)** one defective, **c)** more than one defective.

5) It is known that 10% of resistors produced are defective. What is the probability that in a sample of 12 such resistors more than 2 will be defective?

6) A section in a certain factory consists of 30 work people. Over a long period of time the average number of absentees per shift is 2. Find the probability of more than 2 absentees in any single shift.

7) In a mass production process the average number of defectives produced is 5%. 500 samples each consisting of 20 articles are examined. Show that the expected number of defectives is as shown:

Number of defectives	0	1	2	3	4	5
Frequency	180	189	94	31	4	5

8) In a large batch of fuses it is known that 10% are faulty. Show that in a batch of 500 samples each containing 8 fuses the expected frequencies are:

Number of defectives	0	1	2	3	4	5
Frequency	215	191	75	17	2	0

9) Samples of 6 items are drawn from a production process in which it is known that 10% of the articles made are defective. Draw a histogram showing the probabilities of 0, 1, 2, 3, 4, 5 and 6 defectives in a sample.

THE POISSON DISTRIBUTION

In all cases of repeated trials the binomial theorem may be used to calculate the probabilities. However, as the value of n (the sample size) increases the amount of work involved in calculating the probabilities becomes increasingly large.

In most sampling schemes used in industry, a high degree of accuracy in calculating the probabilities is not essential and approximations to the binomial distribution are therefore used.

One of the approximations used is the Poisson distribution. This is used when:

(a) the probability of the event occurring in a single trial is small, i.e. not greater than 0.1,
(b) the number in the sample is greater than 30,
(c) the expectation, i.e. np, remains constant and is less than 5.

These conditions are almost always satisfied in practical industrial situations. Now there exists a series for e^λ which is:

$$e^\lambda = 1 + \lambda + \frac{\lambda^2}{2!} + \frac{\lambda^3}{3!} + \dots$$

where e is the base of natural logarithms.
The value of e, correct to 4 places of decimals, is 2.7183

In any calculation involving probabilities it is imperative that the total probability, covering all possible events, is equal to 1.

Now $$e^\lambda \times e^{-\lambda} = e^0 = 1$$

Hence we can use the product $e^\lambda \times e^{-\lambda}$ to form a theoretical frequency distribution when it is written:

$$e^{-\lambda}\left(1 + \lambda + \frac{\lambda^2}{2!} + \frac{\lambda^3}{3!} + \dots\right)$$

A distribution obtained by using this series is called a *Poisson distribution* and it forms a close approximation to the binomial distribution provided the conditions given above apply and provided $\lambda = np$.

The successive terms of the series give the probabilities of there being 0, 1, 2, ... defectives in the sample. Thus:

Number of defective items in the sample	Probability of finding the stated number of defective items in the sample
0	$e^{-\lambda}$
1	$(e^{-\lambda})\lambda$
2	$e^{-\lambda}\left(\dfrac{\lambda^2}{2!}\right)$
3	$e^{-\lambda}\left(\dfrac{\lambda^3}{3!}\right)$ etc.

Values of $e^{-\lambda}$ may be obtained using a scientific calculator.

EXAMPLE 21.12

A machine is known to produce 5% of faulty articles. A sample of 4 items is drawn at random from a large batch of these articles. Find the probabilities of obtaining 0, 1, 2, 3 and 4 defectives in the sample using: **a)** the binomial distribution, **b)** the Poisson distribution.

a) For the binomial distribution, $p = 0.05$, $q = 0.95$ and $n = 4$. The expansion of $(q + p)^4$ is,

$$q^4 + 4q^3p + 6q^2p^2 + 4qp^3 + p^4$$

Number of defectives	Term of the expansion	Probability
0	q^4	$(0.95)^4 = 0.814\,51$
1	$4q^3p$	$4 \times (0.95)^3 \times (0.05) = 0.171\,48$
2	$6q^2p^2$	$6 \times (0.95)^2 \times (0.05)^2 = 0.013\,54$
3	$4qp^3$	$4 \times (0.95) \times (0.05)^3 = 0.000\,46$
4	p^4	$(0.05)^4 = 0.000\,01$

Total probability covering all possible cases $= 1.000\,00$

b) For the Poisson distribution, $p = 0.05$, $n = 4$ and $\lambda = np = 0.2$

$$e^{-\lambda} = e^{-0.2} = 0.8187$$

Number of defectives	Term in the series	Probability	
0	$e^{-\lambda}$		0.8187
1	$(e^{-\lambda})\lambda$	0.8187×0.2	$= 0.1637$
2	$e^{-\lambda}\left(\dfrac{\lambda^2}{2!}\right)$	$0.8187 \times \dfrac{(0.2)^2}{2 \times 1}$	$= 0.0164$
3	$e^{-\lambda}\left(\dfrac{\lambda^3}{3!}\right)$	$0.8187 \times \dfrac{(0.2)^3}{3 \times 2 \times 1}$	$= 0.0012$
4	$e^{-\lambda}\left(\dfrac{\lambda^4}{4!}\right)$	$0.8187 \times \dfrac{(0.2)^4}{4 \times 3 \times 2 \times 1}$	$= 0.0000$

Total probability covering all possible cases $= 1.0000$

Comparing the results given by the binomial and Poisson distributions we see that there small discrepancies in the probabilities as given by the two distributions. For most applications these are too small to be significant and we can use the Poisson distribution as an approximation to the binomial distribution *provided* we adhere to the conditions given previously.

EXAMPLE 21.13

A process is known to produce 5% of faulty articles. A sample of 40 such articles is taken. Find: **a)** the chance of obtaining exactly 4 defectives in the sample, **b)** the chance of obtaining more than 4 defectives in the sample. **c)** Represent the probabilities of there being 0, 1, 2, 3 and 4 faulty articles in a histogram.

Here $p = 0.05$ and $\lambda = np = 0.05 \times 40 = 2$

$$e^{-\lambda} = e^{-2} = 0.1353$$

Number of defectives	Term in the series	Probability	
0	$e^{-\lambda}$		0.1353
1	$e^{-\lambda}(\lambda)$	0.1353×2	$= 0.2706$
2	$e^{-\lambda}\left(\dfrac{\lambda^2}{2!}\right)$	$0.1353 \times \dfrac{2^2}{2 \times 1}$	$= 0.2706$
3	$e^{-\lambda}\left(\dfrac{\lambda^3}{3!}\right)$	$0.1353 \times \dfrac{2^3}{3 \times 2 \times 1}$	$= 0.1804$
4	$e^{-\lambda}\left(\dfrac{\lambda^4}{4!}\right)$	$0.1353 \times \dfrac{2^4}{4 \times 3 \times 2 \times 1}$	$= 0.0902$
		Total	$= 0.9471$

a) The chance of obtaining 4 defectives is 0.0902
b) Since the total probability is unity and the sum of the probabilities of obtaining 0, 1, 2, 3 and 4 defectives is 0.9471, the probability of obtaining more than four defectives is $1 - 0.9471 = 0.0529$

c) The histogram is shown in Fig. 21.5.

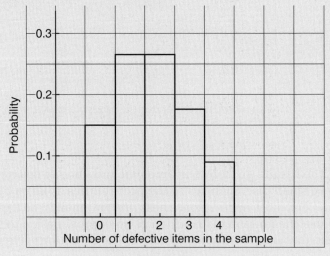

Fig. 21.5

EXAMPLE 21.14

In the mass production of an article 1000 samples each of 50 articles are examined. The numbers of defectives are shown in the following frequency table. Find the average number of defectives per sample and hence show that the distribution is approximately the same as the Poisson distribution with this average.

Defectives	0	1	2	3	4	5
Frequency	674	261	57	7	1	0

$$\text{Total number of defectives} = (261 \times 1) + (57 \times 2) + (7 \times 3) + (1 \times 4)$$
$$= 261 + 114 + 21 + 4$$
$$= 400$$
$$\text{Total number of articles} = 50 \times (674 + 261 + 57 + 7 + 1)$$
$$= 50 \times 1000 = 50\,000$$

$\therefore$
$$p = \frac{400}{50\,000} = 0.008$$

$\therefore$
$$\lambda = np = 50 \times 0.008 = 0.4$$

$\therefore$
$$e^{-\lambda} = e^{-0.4} = 0.6703$$

Number of defectives	Term in the series	Probability		Frequency
0	$e^{-\lambda}$		$= 0.6703$	670.3
1	$(e^{-\lambda})\lambda$	0.6703×0.4	$= 0.2681$	268.1
2	$e^{-\lambda}\left(\dfrac{\lambda^2}{2!}\right)$	$0.6703 \times \dfrac{(0.4)^2}{2 \times 1}$	$= 0.0536$	53.6
3	$e^{-\lambda}\left(\dfrac{\lambda^3}{3!}\right)$	$0.6703 \times \dfrac{(0.4)^3}{3 \times 2 \times 1}$	$= 0.0071$	7.1
4	$e^{-\lambda}\left(\dfrac{\lambda^4}{4!}\right)$	$0.6703 \times \dfrac{(0.4)^4}{4 \times 3 \times 2 \times 1}$	$= 0.0007$	0.7
5	$e^{-\lambda}\left(\dfrac{\lambda^5}{5!}\right)$	$0.6703 \times \dfrac{(0.4)^5}{5 \times 4 \times 3 \times 2 \times 1}$	$= 0.0000$	0

The comparison between the values calculated by the Poisson distribution agree very well with those given in the frequency table.

THE POISSON DISTRIBUTION IN ITS OWN RIGHT

We have seen that we may use the Poisson distribution as an approximation to the binomial distribution. We can do this provided we know the value of n (the sample size) and p (the fraction defective) in the expression $(q + p)^n$. There are, however, very many cases where n is not known. For instance, in checking the number of surface flaws on sheets of plastic or in checking the number of defective items in batches of unknown size.

In such cases we may use the Poisson distribution to calculate the probabilities provided that λ is made equal to the average value of the occurrence of the event.

EXAMPLE 21.15

By checking several boxes containing large numbers of fuses it was found that the average number of defective fuses per box was 3. Find the probability of finding a box containing 4 or more defective fuses.

Here we have $\lambda = 3$ and $e^{-3} = 0.0498$

Number of defective fuses in the sample	Probability	
0	$e^{-\lambda}$	$= 0.0498$
1	$(^-e^{\lambda})\lambda = 0.0498 \times 3$	$= 0.1494$
2	$e^{-\lambda}\left(\dfrac{\lambda^2}{2!}\right) = 0.0498 \times \dfrac{3^2}{2 \times 1}$	$= 0.2241$
3	$(e^{-\lambda})\left(\dfrac{\lambda^3}{3!}\right) = 0.0498 \times \dfrac{3^3}{3 \times 2 \times 1}$	$= 0.2241$
		0.6474

Hence the probability of finding a box containing 4 or more defective fuses is $\qquad p = 1 - 0.6474 = 0.3526$ $\qquad$ We mean that 35.26% of the boxes are likely to contain 4 or more defective fuses.

EXAMPLE 21.16

20 sheets of aluminium alloy were examined for surface flaws. The number of flaws per sheet were as follows:

Sheet number	1 2 3 4 5 6 7 8 9 10 11 12 13 14 15 16 17 18 19 20
Number of flaws	2 0 1 3 2 1 0 0 1 0 2 2 1 0 2 2 1 0 3 2

Find the probability of finding a sheet, chosen at random from a large batch, which has 2 or more surface flaws.

Now $\qquad \lambda = \dfrac{\text{total number of flaws in the 20 sheets}}{20}$

or $\qquad \lambda = \dfrac{25}{20} = 1.25$

$\therefore \qquad e^{-\lambda} = 0.2865$

Number of flaws per sheet	Probability
0	$e^{-\lambda} = 0.2865$
1	$(e^{-\lambda})\lambda = 0.3581$
	$\overline{\quad0.6446\quad}$

Hence the probability of finding a sheet having 2 or more surface flaws is
$p = 1 - 0.6446 = 0.3554$
This probability indicates that it is likely that 35.5% of the sheets will
have 2 or more surface flaws.

Exercise 21.4

1) It is known that 10% of the items produced on a certain machine are defective. Using the Poisson distribution calculate the probabilities of obtaining 0, 1, 2, 3 or 4 defective items in a sample of 4 items.

2) 1% of articles produced in a mass production process are known to be defective. If these are packed in boxes containing 300 articles find the probability that the box will contain exactly 3 defective articles.

3) In a large consignment of resistors the average number in a box that were faulty was 2. Calculate the probability of:
a) finding a box containing exactly 2 defective resistors,
b) finding a box containing 2 or more defective resistors.

4) From the production line of a certain factory 10 successive samples each containing 100 items were checked. The number of items failing the test were: 1, 2, 2, 4, 2, 1, 0, 1, 2, 0. Estimate the probability that a sample of 100 items taken at random will contain 2 or more defective items.

5) 10 sheets of plastic were examined for flaws. The number of flaws per sheet were as follows:

Sheet number	1	2	3	4	5	6	7	8	9	10
No. of flaws	5	3	2	6	4	3	5	2	5	5

Calculate the percentage of sheets which are likely to have 4 or more flaws.

6) Over a long period of time, records show that on the average 3 people in a small factory are absent from a certain shift. Find the probability that: **a)** everyone will be present on the shift,
b) less than 2 will be absent, **c)** 3 or more will be absent.

7) A class of 15 students is prepared for an examination. Experience shows that 10% will obtain a distinction. What is the probability that 1 or more of this class will obtain a distinction in the examination.

8) A pipe line in a chemical processing factory contains a great many valves. It is known that, on average, 2 of these valves will leak. What is the probability that at any given time less than 2 valves will leak?

22 *The Normal Distribution*

The normal distribution curve – mean – spread or dispersion: standard deviation, variance, range – probability from the normal curve – the normal distribution approximating to the binomial distribution for repeated trials – checking for normal distribution.

INTRODUCTION

Normal distribution curves are those we expect to find as a result of taking samples of variate sizes and their frequency of occurrence.

This is the analysis of facts rather than the probability predictions of the previous chapter as to what we think may happen in the future.

You will be introduced to the different techniques used to analyse the results and how they are applied in practice.

REPRESENTATION OF FREQUENCY DISTRIBUTIONS

There are two ways whereby similar articles produced by a process may be checked. They are:

(a) By using limit gauges which merely classify the articles as good or defective.
(b) By measuring the characteristics of the article (e.g. lengths, diameters, etc.) to see if they conform to the drawing specification.

It will be recalled that a frequency distribution may be represented by a frequency curve. If limit gauges are used then the frequency distribution will approximate to a Poisson distribution and the frequency curve will be skewed as shown in Fig. 22.1.

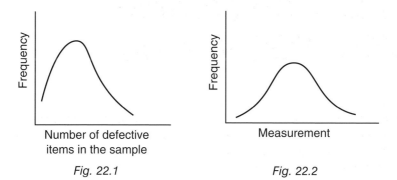

Fig. 22.1   Fig. 22.2

The Normal Distribution

When the data is obtained by actual measurement the frequency curve approximates to the symmetrical bell shaped curve shown in Fig. 22.2. This will only be so if sufficient measurements are made and usually 100 will suffice.

This bell shaped curve is called the *normal distribution curve* and it can be defined in terms of the total frequency, the arithmetic mean and the standard deviation.

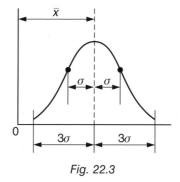

Fig. 22.3

The standard deviation, σ, gives a measure of the spread (or dispersion) of the curve about the mean. As shown in Fig. 22.3, the standard deviation is the distance from the mean to the points of inflexion (points where the curve bends back on itself) of the curve. Although the normal curve extends to infinity on either side of the mean, for most practical purposes it may be assumed to terminate at three standard deviations on either side of the mean.

Since the normal curve is symmetrical about its vertical centre-line, then this centre-line represents the mean of the distribution. This mean locates the position of the curve from the reference axis as shown in Fig. 22.4 which displays similar distributions with different means.

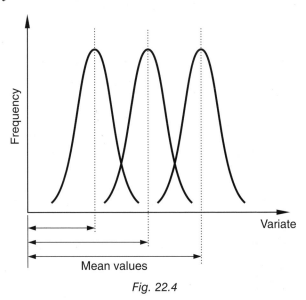

Fig. 22.4

The variance is the square of the standard deviation (σ^2) and this value also helps to describe the spread mathematically.

Although the normal curve extends to infinity on either side of the mean, for most practical purposes it may be assumed to terminate at three standard deviations on either side of the mean (Fig. 22.6).

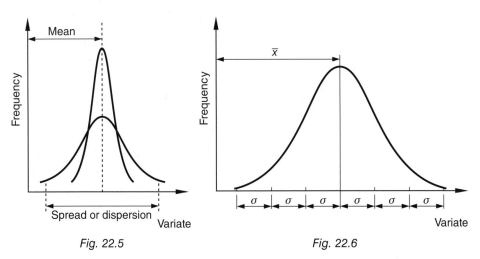

Fig. 22.5   *Fig. 22.6*

Spread or Dispersion

Dispersion is how a distribution is spread out on either side of its central position. Tall and closely packed distributions such as in Fig. 22.4 will have a small dispersion value whilst a low broader figure such as Fig. 22.6 will have a larger value.

The range is the difference between the largest value and the smallest value of a set. It gives some idea of the spread but it depends solely on end values – it has the advantage of being easily obtained, but gives no idea of the distribution of data and is never used as a measure for calculation purposes.

The standard deviation is the most valuable and widely used measure and gives an idea of dispersion about the mean. It is always represented by the Greek letter σ (sigma).

Calculation of the Mean and the Standard Deviation

The formulae used are

$$\text{Mean } \bar{x} = \frac{\Sigma x f}{\Sigma f} \qquad \text{Standard deviation } \sigma = \sqrt{\frac{\Sigma x^2 f}{\Sigma f} - (\bar{x})^2}$$

A calculator will generally have the facilities for you to enter the values of the variate x and the corresponding values of frequency f. The results may then be found from the keys labelled $\bar{x}$ and σ.

Some machines also have another standard deviation labelled $\sigma_n - 1$ which is based on a slightly different formula – this is not the one we shall use.

Each make of calculator will require a different procedure, which will be given in the instruction booklet that accompanies the machine.

However you should be warned that, from experience, it is not easy to enter two sets of numbers accurately. Often the whole procedure has to be gone through **at least** twice to guarantee a correct answer.

Alternatively, the two examples which follow show a tabulation method. This is called the coded method and this uses the idea of choosing a mean value and then working out by how much the true mean value varies.

EXAMPLE 22.1

Calculate the mean and standard deviation for the following frequency distribution.

Resistance (ohms)	5.37	5.38	5.39	5.40	5.41	5.42	5.43	5.44
Frequency	4	10	14	24	34	18	10	6

Our chosen mean value of $x = 5.40$ and unit size $= 0.01$ ohm.

A unit size of 0.01 ohm has been chosen because each value of x differs from its preceding value by 0.01 Making the unit size as large as possible simplifies the calculations.

x	x_c	f	$x_c f$	$x_c^2 f$
5.37	-3	4	-12	36
5.38	-2	10	-20	40
5.39	-1	14	-14	14
5.40	0	24	0	0
5.41	1	34	34	34
5.42	2	18	36	72
5.43	3	10	30	90
5.44	4	6	24	96
	Totals =	120	78	382

Now
$$\bar{x}_c = \frac{\Sigma x_c f}{\Sigma f} = \frac{78}{120} = 0.65$$

$\therefore$ mean $\quad \bar{x} = 5.40 + 0.65 \times 0.01 = 5.4065$ ohm

and
$$\sigma_c = \sqrt{\frac{\Sigma x_c^2 f}{\Sigma f} - (\bar{x}_c)^2} = \sqrt{\frac{382}{120} - (0.65)^2} = 1.662 \text{ ohm}$$

$\therefore$ std. dev. $\quad \sigma = \sigma_c \times \text{unit size} = 1.662 \times 0.01 = 0.016\,62$ ohm

Rough Check on Standard Deviation using the Range

In the last example: $\qquad$ Range $= 5.44 - 5.37 = 0.07$

Hence: $\qquad$ approximately $\sigma = \dfrac{0.07}{6} = 0.012$ ohm

This does not verify the accuracy of the calculated value 0.016 62 but it does show it is of the right order (i.e. not wildly incorrect).

EXAMPLE 22.2

The table indicates experimental results from a sample of resistors.

Resistance (ohm)	24.92–24.94	24.95–24.97	24.98–25.00
Frequency	2	3	9

Resistance (ohm)	25.01–25.03	25.04–25.06	25.07–25.09
Frequency	23	18	5

Calculate the standard deviation and the variance.

The resistance data has been grouped into classes. Here we use the centre of each class for the x value.

Our chosen mean value of $x = 25.02$ ohm and unit size $= 0.03$ ohm.

A unit size of 0.03 ohm has been chosen because each value of x differs from its preceding value by 0.03

Class	x	x_c	f	$x_c f$	$x_c^2 f$
24.92–24.94	24.93	-3	2	-6	18
24.95–24.97	24.96	-2	3	-6	12
24.98–25.00	24.99	-1	9	-9	9
25.01–25.03	25.02	0	23	0	0
25.04–25.06	25.05	1	18	18	18
25.07–25.09	25.08	2	5	10	20
		Totals =	60	7	77

Now $$\bar{x}_c = \frac{\Sigma x_c f}{\Sigma f} = \frac{+7}{60} = +0.116$$

$\therefore$ mean $\quad \bar{x} = 25.02 + (0.03)(0.116) = 25.0235$ ohm

Also $$\sigma_c = \sqrt{\frac{\Sigma x_c^2 f}{\Sigma f} - (\bar{x}_c)^2} = \sqrt{\frac{77}{60} - (0.116)^2} = 1.1268 \text{ ohm}$$

$\therefore$ std. dev. $\quad \sigma = 1.1268 \times 0.03 = 0.0338$ ohm

A rough check gives $\sigma = \dfrac{\text{range}}{6} = \dfrac{25.09 - 24.92}{6} = 0.0283$ ohm

which is of the same order as the value calculated above.

Thus $\quad$ the variance $= \sigma^2 = 0.0338^2 = 0.001\,14$

PROBABILITY FROM THE NORMAL CURVE

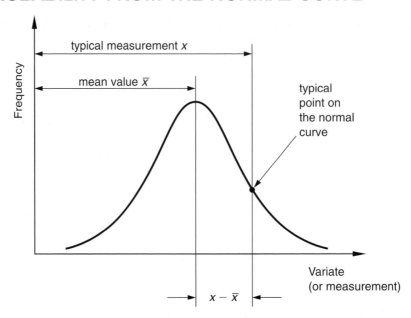

In order for us to make use of the curve it must be 'normalised'. This involves doing three things:

1) Moving the origin to the intersection of the mean ordinate and the horizontal axis on which the variate (or measurements) are plotted.

2) Since the total probability must always be equal to unity, then we make the total area under the curve equal to 1.

3) The horizontal axis is marked off in units of the standard deviation. Any deviation from the origin of a point having measurement x may then be calculated using the expression $u = \dfrac{x - \bar{x}}{\sigma}$.

EXAMPLE 22.3

For a certain process it is found that $\bar{x} = 15.00$ mm and $\sigma = 0.30$ mm.
Express the measurements: **a)** 14.70 mm and
b) 15.21 mm in terms of the standard deviation.

a) Here $x = 14.70$ mm. $u = \dfrac{14.70 - 15.00}{0.30} = -1$

Hence the measurement of 14.70 mm lies 1 standard deviation below
the mean.

Note that when x is less than $\bar{x}$ the value of u is negative.

b) Here $x = 15.21$ mm. $u = \dfrac{15.21 - 15.00}{0.30} = 0.7$

Hence the measurement 15.21 mm lies 0.7 of a standard deviation
above the mean.

Note that when x is greater than $\bar{x}$ the value of u is positive.

Area under the Normal Curve

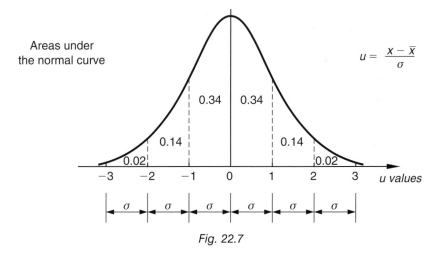

$$u = \frac{x - \bar{x}}{\sigma}$$

Fig. 22.7

Figure 22.7 shows the area of each vertical strip as a percentage of the
total area under the normal curve. These areas may be used to estimate
the chances of something happening which is extremely useful, for
example, in quality control.

For instance there is a chance of 34% of something happening between
the mean and one standard deviation from the mean.

Remember since the total area is 1 then an area of 0.34 represents 34%
of the total area.

The following example will show you how this works.

EXAMPLE 22.4

By measuring a large number of components produced on an automatic lathe it was found that the mean length was 20.10 mm with a standard deviation of 0.03 mm. Find:

a) Within what limits would you expect the lengths for the whole of the components to lie.

b) The chance that one component taken at random would have
    (i) a length greater than 20.16 mm,
    (ii) a length less than 20.07 mm,
    (iii) a length between 20.04 mm and 20.13 mm.

a) For most practical purposes, the normal curve may be regarded as terminating at three standard deviations either side of the mean. Thus, we would expect the lengths for the whole of the components to lie between

$$\text{Mean} \pm 3\sigma = 20.10 \pm 3 \times 0.03 = 20.10 \pm 0.09$$
$$= \text{between } 20.01 \text{ and } 20.19 \text{ mm}$$

b)   (i) For $x = 20.16$ $u = \dfrac{20.16 - 20.10}{0.03} = 2$

Now the area is 2% between 2 and the upper limit, so we would expect 2% of all components to have lengths greater than 20.16 mm.

  (ii) For $x = 20.07$ $u = \dfrac{20.07 - 20.10}{0.03} = -1$

Now the area between -1 and the lower limit is $2\% + 14\% = 16\%$. Thus we would expect 16% of all components to have lengths less than 20.07 mm.

  (iii) For $x = 20.04$ $u = \dfrac{20.04 - 20.10}{0.03} = -2$

       and $x = 20.13$ $u = \dfrac{20.13 - 20.10}{0.03} = 1$

Now the area between -2 and 1 is $14\% + 34\% + 34\% = 82\%$ so we would expect 82% of all components to have lengths between 20.04 and 20.13 mm.

Exercise 22.1

1) In a water absorption test on 100 bricks the following results were obtained:

% absorption	7	8	9	10	11	12	13	14
Frequency	1	4	9	24	30	26	5	1

Calculate the mean and standard deviation.

2) Find the mean and standard deviation for the following distribution which relates to the strength of load carrying bricks:

Strength (N/mm²)	11.46	11.47	11.48	11.49	11.50	11.51	11.52	11.53
Frequency	1	4	12	15	11	6	3	1

3) 100 watt is the nominal value of the sample of electric light bulbs given below. Calculate the mean and standard deviation from the sample.

Power (watt)	99.6	99.7	99.8	99.9	100.0	100.1	100.2	100.3
Frequency	3	8	13	18	15	9	6	3

4) A brand of washing powder tested prior to marketing revealed its capacity to launder woollen garments of equivalent sizes.

Laundered garments	5–7	8–10	11–13	14–16	17–19
Frequency	1	5	11	7	3

From the data above, calculate the mean and standard deviation.

5) It is considered that a person uses, on average, 200 litres of water daily. Using the following data calculate the mean water consumption and the standard deviation.

Daily consumption (litres)	135–155	158–178	181–201	204–224	227–247	250–270
Number of users	3	19	43	26	7	2

6) Measurement from a large batch of mass produced components showed a mean diameter of 18.60 mm, with a standard deviation of 0.02 mm. Find:
   a) within what limits the diameters of the whole of the components would be expected to lie,
   b) the chance that one component taken at random would have:
   (i) a diameter greater than 18.62 mm
   (ii) a diameter less than 18.56 mm.

7) In mass production of bushes it was found that the average bore was 12.5 mm with a standard deviation of 0.015 mm. If 2000 bushes are produced find the number of bushes that are expected to have dimensions between 12.47 and 12.53 mm.

AREAS UNDER THE STANDARD NORMAL CURVE FROM 0 TO u

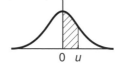

TABLE 22.1

u	0	1	2	3	4	5	6	7	8	9
0.0	.0000	.0040	.0080	.0120	.0160	.0199	.0239	.0279	.0319	.0359
0.1	.0398	.0438	.0478	.0517	.0557	.0596	.0636	.0675	.0714	.0754
0.2	.0793	.0832	.0871	.0910	.0948	.0987	.1026	.1064	.1103	.1141
0.3	.1179	.1217	.1255	.1293	.1331	.1368	.1406	.1443	.1480	.1517
0.4	.1554	.1591	.1628	.1664	.1700	.1736	.1772	.1808	.1844	.1879
0.5	.1915	.1950	.1985	.2019	.2054	.2088	.2123	.2157	.2190	.2224
0.6	.2258	.2291	.2324	.2357	.2389	.2422	.2454	.2486	.2518	.2549
0.7	.2580	.2612	.2642	.2673	.2704	.2734	.2764	.2794	.2823	.2852
0.8	.2881	.2910	.2939	.2967	.2996	.3023	.3051	.3078	.3106	.3133
0.9	.3159	.3186	.3212	.3238	.3264	.3289	.3315	.3340	.3365	.3389
1.0	.3413	.3438	.3461	.3485	.3508	.3531	.3554	.3577	.3599	.3621
1.1	.3643	.3665	.3686	.3708	.3729	.3749	.3770	.3790	.3810	.3830
1.2	.3849	.3869	.3888	.3907	.3925	.3944	.3962	.3980	.3997	.4015
1.3	.4032	.4049	.4066	.4082	.4099	.4115	.4131	.4147	.4162	.4177
1.4	.4192	.4207	.4222	.4236	.4251	.4265	.4279	.4292	.4306	.4319
1.5	.4332	.4345	.4357	.4370	.4382	.4394	.4406	.4418	.4429	.4441
1.6	.4452	.4463	.4474	.4484	.4495	.4505	.4515	.4525	.4535	.4545
1.7	.4554	.4564	.4573	.4582	.4591	.4599	.4608	.4616	.4625	.4633
1.8	.4641	.4649	.4656	.4664	.4671	.4678	.4686	.4693	.4699	.4706
1.9	.4713	.4719	.4726	.4732	.4738	.4744	.4750	.4756	.4761	.4767
2.0	.4472	.4778	.4783	.4788	.4793	.4798	.4803	.4808	.4812	.4817
2.1	.4821	.4826	.4830	.4834	.4838	.4842	.4846	.4850	.4854	.4857
2.2	.4861	.4864	.4868	.4871	.4875	.4878	.4881	.4884	.4887	.4890
2.3	.4893	.4896	.4898	.4901	.4904	.4906	.4909	.4911	.4913	.4916
2.4	.4918	.4920	.4922	.4925	.4927	.4929	.4931	.4932	.4934	.4936
2.5	.4938	.4940	.4941	.4943	.4945	.4946	.4948	.4949	.4951	.4952
2.6	.4953	.4955	.4956	.4957	.4959	.4960	.4961	.4962	.4963	.4964
2.7	.4965	.4966	.4967	.4968	.4969	.4970	.4971	.4972	.4973	.4974
2.8	.4974	.4975	.4976	.4977	.4977	.4978	.4979	.4979	.4980	.9481
2.9	.4981	.4982	.4982	.4983	.4984	.4984	.4985	.4985	.4986	.4986
3.0	.4987	.4987	.4987	.4988	.4988	.4989	.4989	.4989	.4990	.4990
3.1	.4490	.4991	.4991	.4991	.4992	.4992	.4992	.4992	.4993	.4993
3.2	.4993	.4993	.4994	.4994	.4994	.4994	.4994	.4995	.4995	.4995
3.3	.4995	.4995	.4995	.4996	.4996	.4996	.4996	.4996	.4996	.4997
3.4	.4997	.4997	.4997	.4997	.4997	.4997	.4997	.4997	.4997	.4998
3.5	.4998	.4998	.4998	.4998	.4998	.4998	.4998	.4998	.4998	.4998
3.6	.4998	.4998	.4999	.4999	.4999	.4999	.4999	.4999	.4999	.4999
3.7	.4999	.4999	.4999	.4999	.4999	.4999	.4999	.4999	.4999	.4999
3.8	.4999	.4999	.4999	.4999	.4999	.4999	.4999	.4999	.4999	.4999
3.9	.5000	.5000	.5000	.5000	.5000	.5000	.5000	.5000	.5000	.5000

USING THE TABLE OF AREAS (TABLE 22.1)

Figure 22.8 shows a typical area. To use the table first note that the figures in the first column are values of u in increments of 0.1. The corresponding value in the column headed 0 gives the area between $u = 0$ and the value of u in column one.

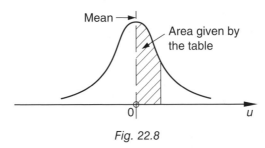

Fig. 22.8

Thus the area between $u = 0$ and $u = 0.7$ is 0.2580
If the value of u has two decimal places the area will be found in the appropriate column of the next nine columns.
Thus the area between $u = 0$ and $u = 0.74$ is 0.2704

If the value of u is negative, the table is used in exactly the same way and in reading the table the negative sign is ignored.

The following examples will show how the table is used:

EXAMPLE 22.5

Find the area under the normal curve between:

a) $u = 0$ and $u = 0.63$ **b)** $u = 0$ and $u = -0.85$
c) $u = 1.2$ and $u = 1.73$ **d)** $u = -0.63$ and $u = 1.12$

a) The area between $u = 0$ and $u = 0.63$ is as shown in Fig. 22.9 and is found directly from the table as 0.2357

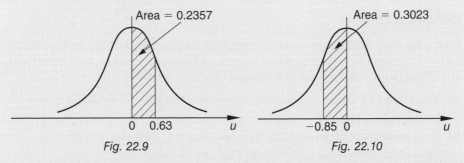

Fig. 22.9   Fig. 22.10

b) Since the normal curve is symmetrical about the mean we can find the required area between $u = 0$ and $u = -0.85$ directly from the table as if it was that between $u = 0$ and $u = 0.85$. Its value is 0.3023 (see Fig. 22.10).

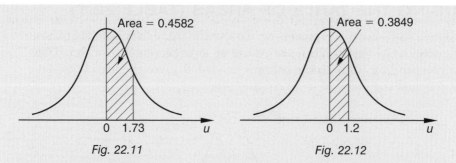

Fig. 22.11   *Fig. 22.12*

c) First find the area between $u = 0$ and $u = 1.73$. This is 0.4582 and is shown in Fig. 22.11. Next find the area between $u = 0$ and $u = 1.2$ This is 0.3849 and is shown in Fig. 22.12.

By subtracting the area 0.3849 from the area 0.4582 we obtain the required area, 0.0733, which is shown in Fig. 22.13.

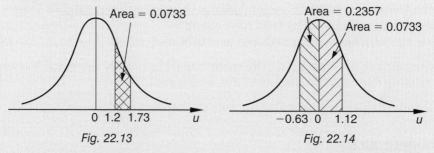

Fig. 22.13   *Fig. 22.14*

d) First we find the area between $u = 0$ and $u = -0.63$, which is the same as the area between $u = 0$ and $u = 0.63$. Its value is 0.2357. Next find the area between $u = 0$ and $u = 1.12$, which is 0.3686

The required area is the sum of these two, as shown in Fig. 22.14, giving a value of 0.6043

EXAMPLE 22.6

By measuring a large number of components produced on an automatic lathe it was found that the mean length of the components was 20.10 mm with a standard deviation of 0.03 mm. Find:

a) Within what limits you would expect the lengths for the whole of the components to lie.
b) The probability that one component taken at random would have:
   (i) a length between 20.05 mm and 20.12 mm,
   (ii) a length less than 20.02 mm, (iii) a length greater than 20.17 mm.

a) As previously explained, for most practical purposes the normal curve may be regarded as terminating at 3 standard deviations on either side of the mean. Thus, we would expect the lengths for the whole of the component to lie between:

$$20.10 \pm 3\sigma = 20.10 \pm 0.03 = 20.10 \pm 0.09 \text{ mm}$$

that is between 20.01 mm and 20.19 mm.

b) (i) $u_1 = \dfrac{20.05 - 20.10}{0.03} = -1.67$ $u_2 = \dfrac{20.12 - 20.10}{0.03} = 0.67$

using the table of areas under the normal curve:

Area under the curve between $u = 0$ and $u = -1.67 = 0.4525$

Area under the curve between $u = 0$ and $u = 0.67 = 0.2486$

Hence between $u = 0$ and $u = 0.67$

total area under the curve $= 0.4525 + 0.2486 = 0.7011$

The probability is 0.7011 or 70.11%. This indicates that 70.11% of all the components are expected to have length between 20.05 mm and 20.12 mm.

(ii) $$u = \frac{20.02 - 20.10}{0.03} = -2.67$$

Now between $u = 0$ and $u = -2.67$

the area under curve $= 0.4962$

Since the normal curve is symmetrical and since its area is unity, the total area to the left of the mean is 0.5000. The probability of obtaining a component with a length less than 20.02 mm is therefore $0.5000 - 0.4962 = 0.0038$ or 0.38%. This means that 0.38% of all components produced are expected to have a length less than 20.02 mm.

(iii) $$u = \frac{20.17 - 20.1}{0.03} = 2.33$$

Now between $u = 0$ and $u = 2.33$

the area under curve $= 0.4901$

The probability of obtaining a component with a length greater than 20.17 mm is thus $0.5000 - 0.4901 = 0.0099$ or 0.99%. This means that 0.99% of all the components produced are expected to have a length greater than 20.17 mm.

EXAMPLE 22.7

10 000 resistors are to be produced with a specification limit of 100 ± 7 ohms. By measuring the first 100 made it was found that their mean resistance was 102 ohms with a standard deviation of 5 ohms. Determine how many of these resistors are likely to be outside the specification limits.

The first step is to calculate the value of u at the lower specification limit.

$$u_1 = \frac{93 - 102}{5} = -1.8$$

The area between $u = 0$ and $u = -1.8$ is 0.4641 Hence the tail area to the left of $u = -1.8$ is $0.5 - 0.4641 = 0.0359$ or 3.59% (Fig. 22.15). Hence the number of resistors likely to be produced outside of the lower specification limit is 3.59% of $10\,000 = 359$.

We next calculate the value of u at the upper specification limit.

$$u_2 = \frac{107 - 102}{5} = 1$$

The area between $u = 0$ and $u = 1$ is 0.3413 The tail area to the right of $u = 1$ is $0.5 - 0.3413 = 0.1587$ or 15.87% Hence the number of resistors likely to be produced outside of the upper specification limit is 15.87% of $10\,000 = 1587$.

The total number of resistors likely to be produced outside of specification limits is $359 + 1587 = 1946$.

Note that in Fig. 22.15 it is the shaded tail areas which give the probabilities of a resistor being produced which is outside of the specification limits.

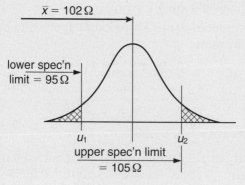

Fig. 22.15

EXAMPLE 22.8

On checking a large number of steel bars it was found that 5% were under 132.7 mm in length. 35% were between 132.7 mm and 135.5 mm. Find the mean and standard deviation assuming a normal distribution.

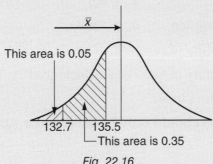

Fig. 22.16

The given percentages are represented as areas in Fig. 22.16. We find that for an area of 0.45, $u = 1.65$, and for an area of 0.10, $u = 0.25$,

and substituting in the equation: $u = \dfrac{x - \bar{x}}{\sigma}$

Then
$$-1.65 = \frac{132.7 - \bar{x}}{\sigma}$$

or
$$-1.65\sigma = 132.7 - \bar{x} \qquad [1]$$

And
$$-0.25 = \frac{135.5 - \bar{x}}{\sigma}$$

or
$$-0.25\sigma = 135.5 - \bar{x} \qquad [2]$$

Subtracting equation [2] from equation [1],
$$-1.40\sigma = -2.8$$

∴
$$\sigma = \frac{2.8}{1.40} = 2.00 \text{ mm}$$

Substituting for σ in equation [2],
$$-0.25 \times 2 = 135.5 - \bar{x}$$

∴
$$-0.5 = 135.5 - \bar{x}$$

∴
$$\bar{x} = 135.5 + 0.5$$
$$\bar{x} = 136.0 \text{ mm}$$

THE NORMAL DISTRIBUTION AS AN APPROXIMATION TO THE BINOMIAL DISTRIBUTION FOR REPEATED TRIALS

It can be shown that, for a binomial distribution:

the mean $\bar{x} = np$

and the standard deviation $\sigma = \sqrt{npq}$

where p = probability of success in a single trial,

q = probability of failure in a single trial = $1 - p$

and n = number of trials.

This approximation should only be used when np is greater than 5

EXAMPLE 22.9

A manufacturer of light bulbs finds that on the average 3% are defective. What is the probability that 40 or more will be defective in 1000 bulbs selected at random?

$$p = 0.03 \quad q = 0.97$$

$$\bar{x} = np = 1000 \times 0.03 = 30$$

$$\therefore \quad \sigma = \sqrt{npq} = \sqrt{1000 \times 0.03 \times 0.97}$$

$$= \sqrt{29.1} = 5.394$$

$$u = \frac{x - \bar{x}}{\alpha} = \frac{40 - 30}{5.394} = \frac{10}{5.394} = 1.85$$

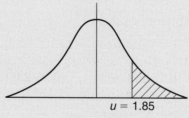

$u = 1.85$

Fig. 22.17

We now need to find the shaded area of Fig. 22.17. The area between $u = 0$ and $u = 1.85$ is 0.4678. The probability is, therefore,

$$0.5000 - 0.4678 = 0.0322 \text{ or } 3.22\%$$

CHECKING A DISTRIBUTION TO SEE IF IT IS A NORMAL DISTRIBUTION

One way of checking a distribution to see if it is a normal distribution is to use probability paper. The method is shown in Example 22.10.

EXAMPLE 22.10

Check if the distribution given below is a normal distribution.

Variate	2.26	2.27	2.28	2.29	2.30	2.31	2.32	2.33	2.34
Frequency	1	3	25	60	126	61	20	3	1

The first step is to convert the given frequency distribution into a cumulative frequency distribution. Remembering that the upper boundary limit of the first class is 2.265, the cumulative frequency distribution is then as follows:

Variate	Cumulative frequency	Percentage cumulative frequency
less than 2.265	1	$\frac{1}{300} \times 100 = 0.33\%$
less than 2.275	4	$\frac{4}{300} \times 100 = 1.33\%$
less than 2.285	29	$\frac{29}{300} \times 100 = 8.17\%$
less than 2.295	89	29.7%
less than 2.305	215	71.7%
less than 2.315	276	90.2%
less than 2.325	296	98.7%
less than 2.335	299	99.7%
less than 2.345	300	100 %

The percentage cumulative frequencies are plotted against the variate on the probability paper shown in Fig. 22.18. The closeness to which the plotted points conform to a straight line determines whether the distribution is nearly normal or not. From Fig. 22.18 we shall see that the given distribution may, for all practical purposes, be accepted as a normal distribution.

The mean and standard deviation may also be found from Fig. 22.18. For a normal distribution the mean is the value of the variate corresponding to a cumulative percentage frequency of 50%. From the diagram the mean of the distribution is found to be about 2.30

For a normal distribution the area under the normal curve between the mean and 1 standard deviation is 0.3413 or 34.13% (see Table 22.1). Now 50% − 34.13% = 15.87%. Value of the variate corresponding to 15.87% = 2.288. Hence the standard deviation is 2.30 − 2.288 = 0.012

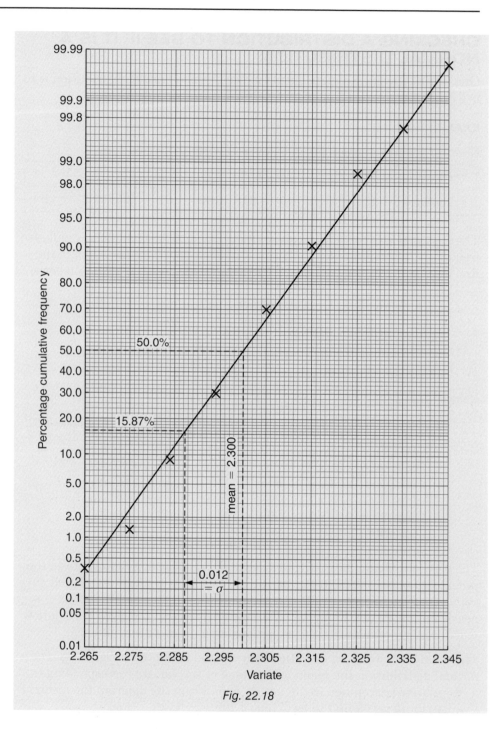

Fig. 22.18

Exercise 22.2

In the following $u = (x - \bar{x})/\sigma$

1) If $\bar{x} = 30$ and $\sigma = 5$ find the values of u corresponding to:

    a) $x = 22$ **b)** $x = 28$ **c)** $x = 35$ **d)** $x = 41$

2) Find the areas under the normal curve:

    a) between $u = 0$ and $u = 1.20$

    b) between $u = -0.75$ and $u = 0$

    c) between $u = -0.3$ and $u = 1.12$

    d) between $u = -2.3$ and $u = -1.56$

    e) between $u = 0.58$ and $u = 2.41$

    f) less than $u = -0.79$

    g) greater than $u = 2.11$

    h) greater than $u = -0.86$

    i) less than $u = 1.12$

    j) less than $u = 2.49$

3) By measuring a large number of components produced on an automatic lathe it was found that the mean length of the components was 18.60 mm with a standard deviation of 0.02 mm. Find:

    a) within what limits the lengths of the whole of the components could be expected to lie,

    b) the probability that one component taken at random would have:
        (i) a length less than 18.56 mm,
        (ii) a length greater than 18.65 mm,
        (iii) a length between 18.59 mm and 18.62 mm.

4) For a normal distribution with a mean of 20 and a standard deviation of 4 find the probabilities for:

    a) the interval 12 to 18,   **b)** the interval 17 to 25,

    c) the interval 15 to 30,   **d)** 28 and above,

    e) 14 and less.

5) In mass producing a bush it is found that the average diameter is 12.5 mm with a standard deviation of 0.015 mm. If 2000 bushes are to be produced find the number of bushes that can be expected to have dimensions between 12.475 and 12.525 mm.

6) An article is being produced on an automatic lathe. 300 of these are measured and a certain dimension is checked to the nearest 0.01 mm. The results are given in the following table:

Dimension	9.96	9.97	9.98	9.99	10.00	10.01	10.02	10.03	10.04
Frequency	1	6	25	72	93	69	27	6	1

Find the probability that one article chosen at random would have:

a) a dimension less than 9.975 mm,

b) a dimension greater than 10.035 mm,

c) a dimension between 9.985 and 10.015 mm.

7) In checking the heights of a large group of men it is found that 8% are under 1.55 m in height and 40% have a height between 1.55 m and 1.68 m. Assuming that a normal distribution holds find the mean and standard deviation of the group, what percentage of the group may be expected to have a height greater than 1.73 m ?

8) The masses of a group of 200 castings are given in the table below:

Mass (kg)	168	168.5	169	169.5	170	170.5	171	171.5	172
Frequency	2	2	28	44	50	38	26	6	4

Find the probability of the mass of a single casting falling between 168.75 kg and 171.25 kg.

9) An article is being mass produced. The drawing demands that a certain dimension shall be 12.51 ± 0.04 mm.
600 of the articles are measured with the following results:

Dimension (mm)	12.46	12.47	12.48	12.49	12.50	12.51	12.52	12.53	12.54
Frequency	1	7	52	121	250	123	40	5	1

Calculate the mean and standard deviation. Hence find how many of the 600 articles are likely to meet the drawing specification.

10) In a certain factory 1000 electric lamps are installed. These lamps have an average burning life of 1100 hours, and a standard deviation of 250 hours. How many lamps are likely to fail during the first 800 burning hours and how many lamps are likely to fail between 900 and 1400 burning hours ?

11) 1000 resistors are to be made with specification limits of 33 ± 0.6 ohm. It is known that the mean of the resistors will be 33.2 ohm with a standard deviation of 0.3 ohm. Find how many of the 1000 resistors will be produced outside of the specification limits.

12) As a result of testing 300 oil filled capacitors the following frequency distribution was obtained.

Capacity (μF)	19.96	19.97	19.98	19.99	20.00	20.01	20.02	20.03	20.04
Frequency	1	6	25	72	93	69	27	6	1

By using probability paper establish if the distribution is approximately normal or not and estimate the mean and standard deviation.

13) The masses of 200 castings were as shown in the following table:

Mass (kg)	168	168.5	169	169.5	170	170.5	171	171.5	172
Frequency	2	2	28	44	50	38	26	6	4

By using probability paper show that the distribution is very nearly normal and hence determine the mean and standard deviation.

14) A manufacturer knows that 2% of his products are, on the average, defective. What is the probability that 1000 articles will contain 18 or more defectives?

15) The average life of a certain battery is 24 months with a standard deviation of 6 months. If 800 batteries are to be sent to a distributor how many can be expected to fail between 12 and 18 months?

16) In a mass production process it is known that 2% of the articles produced are defective. What is the probability of obtaining 22 or more defectives in a sample of 800 such articles?

17) When samples of 1500 articles are examined it is found that on the average 15 articles are defective. In a sample of 1500 articles what is the probability that it will contain between 13 and 18 defectives?

23 Sampling and Estimation Theory

Random sampling – random numbers – sampling distributions of the means of samples – standard error for the distribution of sample means – confidence levels, limits, intervals and coefficients.

INTRODUCTION

We are often faced with the problem of finding parameters, usually the mean and standard deviation, of a large parent population.

One way is to take random samples of the same size and use values obtained from these to estimate the required parameter values.

Having obtained our calculated results we may need to know how accurate are our predictions. We use confidence levels which are based on the normal distribution curve and the areas below the curve on either side on the mean, as used in the previous chapter.

RANDOM SAMPLING

Suppose that from a parent population we take samples each containing the same number of items.

The samples which are chosen must be representative of the population. Representative samples may be obtained by *random sampling* in which each item of the population has an equal chance of being included in the sample. One way of obtaining a random sample is to give each item in the population a number and conduct a raffle. A second way is to use a table of *random numbers*, part of which is shown below, to obtain the sample.

RANDOM NUMBERS

50532	25496	96652	42657	73557	76152
50020	24819	52984	76168	07136	40876
79971	54195	25708	51817	36732	72484

EXAMPLE 23.1

The table below gives the diameters of 100 turned bars produced on an automatic lathe.

Diameter (mm)	Frequency
44.96	8
44.97	21
44.98	43
44.99	20
45.00	8

Three samples, each containing 5 items, are required from this population. Using a table of random numbers show how the samples are obtained.

Using two digits to number each of the 100 bars, the 8 bars of 44.96 mm diameter are numbered 00, 01, 02, 03, 04, 05, 06 and 07. The 21 bars of 44.97 mm diameter are numbered 08 to 28 and so on as shown in the table below.

Diameter (mm)	Frequency	Sampling number
44.96	8	00–07
44.97	21	08–28
44.98	43	29–71
44.99	20	72–91
45.00	8	92–99

Sampling numbers are now taken from the table of random numbers given above. From the first line the sequence 50, 53, 22, 54 and 96 is obtained. Hence the first sample will comprise of bars numbered 50, 53, 22, 54 and 96. Carrying on in this way we find the second sample to comprise of bars numbered 95, 65, 24, 26 and 57. The third sample comprises of bars numbered 73, 55, 77, 61 and 52.

FINITE OR INFINITE POPULATION

A population may be finite or it may be infinite. If we draw 5 balls successively from a box containing 100 balls, without replacing them, we are sampling from a finite population. If, however, we replace each ball after drawing it we are sampling from an infinite population, since theoretically we may draw any number of samples without exhausting the population.

In practice, sampling without replacement from a very large population may be considered as sampling from an infinite population.

SAMPLING DISTRIBUTIONS

If we wish to investigate a population we may take random samples of the same size and then calculate the mean and standard deviation for each sample. These values will differ from sample to sample, and to handle these variations we will group the results together and obtain e.g. a set of all the sample means, or a set of all the sample standard deviations. These sets are called *sampling distributions*.

The **sampling distributions** are approximately **normal**

a) Providing that **each sample contains 30 or more items** irrespective of whether the population itself is a normal distribution

b) If its **standard deviation** σ_p **is known** and if the **population is normally distributed**, even for small numbers of items per sample.

SAMPLING MEAN AND STANDARD DEVIATION (OR STANDARD ERROR)

If all possible samples of size n_s are drawn from **a finite population** (**no replacement**) of size n_p

then   sample mean $\bar{x}_s = \bar{x}_p$

and   sample 'standard deviation' $\sigma_s = \dfrac{\sigma_p}{\sqrt{n_s}} \sqrt{\dfrac{n_p - n_s}{n_p - 1}}$

For a sampling distribution the *standard deviation* is also called the *standard error*.

If the **population is infinite**, or the **sampling is with replacement**, then the results given above reduce to

$$\bar{x}_s = \bar{x}_p \quad \text{and} \quad \sigma_s = \dfrac{\sigma_p}{\sqrt{n_s}}$$

EXAMPLE 23.2

A population consists of the five numbers 1, 2, 3, 4 and 5. Find:

a) the mean and standard deviation of the population,
b) the mean and standard deviation of all the possible samples of size 2 which can be drawn with replacement from this population.

In this problem we have two complete populations (albeit small), in part **a)** the given numbers, and in part **b)** the sample means. For both of these we will find their respective mean and standard deviations using the formulae met previously, namely

$$\text{mean } \bar{x} = \frac{\Sigma xf}{\Sigma f} \qquad \text{standard deviation} \quad \sigma = \sqrt{\frac{\Sigma x^2 f}{\Sigma f} - \bar{x}^2}$$

In both these populations there is only one of each 'item' so

$$\Sigma f = N \quad \text{where } N \text{ is the number of items}$$

and the formulae become:

$$\text{mean } \bar{x} = \frac{\Sigma x}{N} \qquad \text{standard deviation} \quad \sigma = \sqrt{\frac{\Sigma x^2}{N} - \bar{x}^2}$$

a) The mean of the population is:

$$\bar{x}_p = \frac{1 + 2 + 3 + 4 + 5}{5} = \frac{15}{5} = 3$$

The standard deviation of the population is:

$$\sigma_p = \sqrt{\frac{\Sigma x^2}{N} - \bar{x}_p{}^2}$$

$$= \sqrt{\frac{1^2 + 2^2 + 3^2 + 4^2 + 5^2}{5} - 3^2}$$

$$= \sqrt{2} = 1.414$$

b) There are 25 samples of size 2 which can be drawn, with replacement, from the population. They are:

$$
\begin{array}{ccccc}
1,1 & 2,1 & 3,1 & 4,1 & 5,1 \\
1,2 & 2,2 & 3,2 & 4,2 & 5,2 \\
1,3 & 2,3 & 3,3 & 4,3 & 5,3 \\
1,4 & 2,4 & 3,4 & 4,4 & 5,4 \\
1,5 & 2,5 & 3,5 & 4,5 & 5,5
\end{array}
$$

The corresponding sample means (the $\bar{x}_i$ values) are:

$$
\begin{array}{ccccc}
1.0 & 1.5 & 2.0 & 2.5 & 3.0 \\
1.5 & 2.0 & 2.5 & 3.0 & 3.5 \\
2.0 & 2.5 & 3.0 & 3.5 & 4.0 \\
2.5 & 3.0 & 3.5 & 4.0 & 4.5 \\
3.0 & 3.5 & 4.0 & 4.5 & 5.0
\end{array}
$$

The mean of the sampling distribution is:

$$
\bar{x}_s = \frac{\text{sum of all the sample means}}{\text{number of samples}} = \frac{75}{25} = 3
$$

This illustrates the fact that:

$$
\bar{x}_s = \bar{x}_p
$$

The standard deviation of the sampling distribution is:

$$
\sigma_s = \sqrt{\frac{\Sigma \bar{x}_i^2}{N} - \bar{x}_s^2}
$$

where $\quad \bar{x}_i$ = the mean of the ith sample

and $\quad N$ = the number of samples taken

Thus $\quad \sigma_s = \sqrt{\frac{(1.0^2 + 1.5^2 + 2.0^2 + ...)}{25} - 3^2} = 1$

This illustrates the fact that for sampling with replacement:

$$
\sigma_s = \frac{\sigma_p}{\sqrt{n_s}} = \frac{\sqrt{2}}{\sqrt{2}} = 1
$$

EXAMPLE 23.3

The heights of 2000 male factory workers are normally distributed with a mean of 172 cm and a standard deviation of 7.5 cm. If 40 samples each containing 30 workers are taken, determine the mean and standard deviation of the sampling distribution for sample means if:

a) the sampling is done with replacement.
b) the sampling is done without replacement.

a) $\bar{x}_s = \bar{x}_p = 172\,\text{cm}$

$$\sigma_s = \frac{\sigma_p}{\sqrt{n_s}} = \frac{7.5}{\sqrt{30}} = 1.369\,\text{cm}$$

b) $\bar{x}_s = \bar{x}_p = 172\,\text{cm}$

$$\sigma_s = \frac{\sigma_p}{\sqrt{n_s}}\sqrt{\frac{n_p - n_s}{n_p - 1}} = \frac{7.5}{\sqrt{30}}\sqrt{\frac{2000 - 30}{2000 - 1}} = 1.359\,\text{cm}$$

Thus the value of σ_s for sampling done without replacement is only slightly less than that for sampling done with replacement. Hence the population of 2000 is large enough to be considered as infinite population.

A NORMAL SAMPLING DISTRIBUTION DIAGRAM

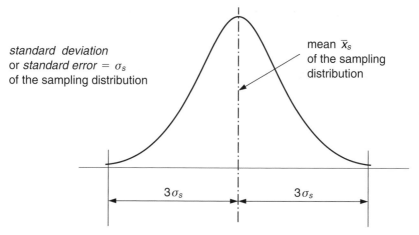

Fig. 23.1   *A normal sampling distribution diagram*

EXAMPLE 23.4

A very large batch of ball bearings has a mean diameter of 25.02 mm with a standard deviation of 0.30 mm. A random sample of 100 of these ball bearings is taken. Find:

a) the probability that the mean of the sample lies between 24.96 mm and 25.00 mm,

b) the probability that the mean of the sample is greater than 25.10 mm.

A very large batch implies a infinite population, so for the sampling distribution of means:

$$x_s = x_p = 25.02 \text{ mm} \qquad \sigma_s = \frac{\sigma_p}{\sqrt{n_s}} = \frac{0.30}{\sqrt{100}} = 0.03 \text{ mm}$$

In addition, since the sample size is 100 (n_s greater than 30) the sampling distribution for the sample means will be a normal curve. We are therefore able to use the table of the areas under the normal curve (Table 22.1 on page 288) to estimate probabilities.

a) When $\bar{x}_1 = 24.96$ mm, $u_1 = \dfrac{\bar{x}_1 - \bar{x}_p}{\sigma_s} = \dfrac{24.96 - 25.02}{0.03} = -2.00$

When $\bar{x}_2 = 25.00$ mm, $u_2 = \dfrac{\bar{x}_2 - \bar{x}_p}{\sigma_s} = \dfrac{25.00 - 25.02}{0.03} = -0.67$

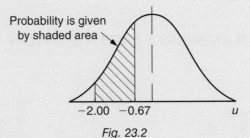

Probability is given by shaded area

−2.00 −0.67 u

Fig. 23.2

Now between $u = -2.00$ and $u = 0$ area $= 0.4472$

and between $u = -0.67$ and $u = 0$ area $= 0.2486$

Thus between $u = -2.00$ and $u = -0.67$ area $= 0.4472 - 0.2486$
$$= 0.1986$$

Hence the probability that the mean of the sample lies between 24.96 mm and 25.00 mm is 0.1986

b) When $\bar{x}_3 = 25.10,$ $u_3 = \dfrac{25.10 - 25.02}{0.03} = 2.67$

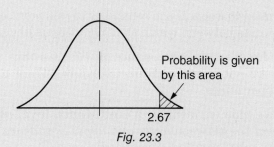

Probability is given by this area

2.67

Fig. 23.3

From Table 22.1 and Fig. 23.3,

Now between $u = 0$ and $u = 2.67$ area = 0.4962

thus beyond $u = 2.67$ tail area = 0.5 − 0.4962
                                  = 0.0038

Hence the probability that the mean of the sample is greater than 25.10 mm is 0.0038

POINT AND INTERVAL ESTIMATES

Using a sampling distribution we may obtain a value of the sample mean – this single number is called a point estimate.

What we really need to know is the reliability of this point estimate. For this we use limits between which the true mean is likely to lie.

Remember how, for example, a resistance of $52 \pm 0.5\,\text{ohm}$ lies between the extremes of $52 - 0.5\,\text{ohm}$ and $52 + 0.5\,\text{ohm}$. These extremes are the lower and upper limits.

Figure 23.4 shows what we mean by **a confidence level of 95%**, and the *confidence interval* the between *lower and upper confidence limits.* The value 1.96 is called *the confidence coefficient.*

Thus for a 95% confidence level the confidence limits of the population mean are given by $\bar{x}_p \pm 1.96\sigma_s$.

The 95% Confidence Interval

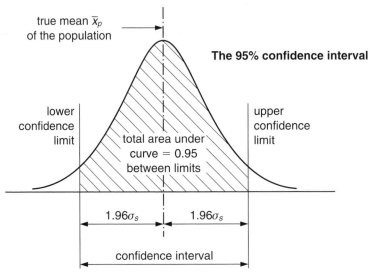

Fig. 23.4   The 95% confidence interval

The most commonly used confidence levels are 90%, 95%, 99% and 99.9%

– the table below shows the relevant limits:

Confidence level	90%	95%	99%	99.9%
Limits	$\bar{x}_p \pm 1.65\sigma_s$	$\bar{x}_p \pm 1.96\sigma_s$	$\bar{x}_p \pm 2.58\sigma_s$	$\bar{x}_p \pm 3.29\sigma_s$

These confidence limits are examples of interval estimates of the population mean because we are stating limits between which the population mean may be expected to lie. Although the sample mean is useful as an estimate of the population mean there is no way by which we can express the accuracy of the estimate. (In fact, mathematically speaking, the probability of the mean of the sample exactly equalling the mean of the population is $p = 0$.) However, by using confidence limits, the degree of accuracy of the estimate is given by the confidence level.

EXAMPLE 23.5

The diameters of a random sample of 100 turned bars were measured and their mean diameter was found to be 18.10 mm. The standard deviation for the population is known to be 1.2 mm. At the 99% confidence level, what is the estimate of the mean diameter of all the turned bars produced?

Since no mention is made of the population size we will assume it to be infinite, and since the sample number n_s is greater than 30 then:

sample mean $\bar{x}_s = \bar{x}_p$ and so the population mean is also 18.10 mm

and $$\sigma_s = \frac{\sigma_p}{\sqrt{n_s}} = \frac{1.2}{\sqrt{100}} = 0.12$$

For a 99% confidence level

$$\text{Limits} = \bar{x}_p \pm 2.58\sigma_s = 18.10 \pm 2.58 \times 0.12$$
$$= 18.10 \pm 0.31$$

So we can be 99% certain that the true population mean of diameters lies between 17.79 mm and 18.41 mm.

EXAMPLE 23.6

A manufacturer produces batches of 2000 small spindles, and over a long period and the standard deviation of their lengths was found to be 0.21 mm. From each batch a random sample of 100 spindles were found to have a mean length of 6.23 mm. Estimate the limits of the lengths in a batch for a confidence level of **a)** 95% **b)** 98%.

The population size $n_p = 2000$ and standard deviation $\sigma_p = 0.21$

The sample size $n_s = 100$ and its mean is $\bar{x}_s = 6.23$

Here we have a finite population so

sample mean $\bar{x}_s = \bar{x}_p$ and so the population mean n_p is also 6.23

and $\qquad \sigma_s = \dfrac{\sigma_p}{\sqrt{n_s}} \sqrt{\dfrac{n_p - n_s}{n_p - 1}} = \dfrac{0.21}{\sqrt{100}} \sqrt{\dfrac{2000 - 100}{2000 - 1}} = 0.0205$

a) For 95% confidence level

Limits $= \bar{x}_p \pm 1.96\sigma_s = 6.23 \pm 1.96 \times 0.0205$

$\qquad\qquad\qquad = 6.23 \pm 0.0402$

So we can be 95% certain that the true population mean of lengths lies between 6.19 mm and 6.27 mm.

b) For 98% confidence level we shall have to use Table 22.1 for areas under the normal curve. The 'half' area will be one half of 0.98 which is 0.49 – the nearest to this in the body of the table is 0.4901 corresponding to a value $u = 2.33$ which is the value of the confidence coefficient we need.

Limits $= \bar{x}_p \pm 2.33\sigma_s = 6.23 \pm 2.33 \times 0.0205$

$\qquad\qquad\qquad = 6.23 \pm 0.0478$

So we can be 98% certain the true population mean of lengths lies between 6.18 mm and 6.28 mm.

It is interesting to note that the higher the confidence level we require then the greater is the interval between the limits.

Exercise 23.1

1) A population consists of the four numbers 2, 6, 10 and 14. Consider all the samples of size two which can be drawn, with replacement, from this population. Calculate:
   a) the population mean,
   b) the population standard deviation,
   c) the mean of the sampling distribution for means,
   d) the standard deviation of the sampling distribution for means,
   e) Verify c) and d) by using suitable formulae.

2) The masses of 3000 ball bearings are normally distributed with a mean of 635 grams and a standard deviation of 1.4 grams. If 70 samples each containing 36 ball bearings are taken from this population, find the expected mean and standard deviation of the distribution of samples if the sampling is done: a) with replacement, b) without replacement.

3) The heights of 100 male factory workers represent a random sample of the heights of all 1972 male workers in the factory. The mean height of the 100 men measured was 1.727 m with a standard deviation of 0.050 m. Determine unbiased estimates of:
   a) the mean height of the population,
   b) the standard deviation of the population.

4) 400 resistors have a mean resistance of 50.2 ohm and a standard deviation of 0.5 ohm. Find the probability that a random sample of 50 of these resistors would have a resistance:
   a) between 50.15 ohm and 50.30 ohm,
   b) less than 50.05 ohm,   c) greater than 50.38 ohm.

5) The diameters of 5000 electric motor shafts are normally distributed with a mean of 50.42 mm and a standard deviation of 1.23 mm. If 80 samples each containing 30 shafts are obtained find the expected mean and standard deviation of the sampling distribution for sample means if the sampling was done:
   a) with replacement,   b) without replacement,
   c) how many samples are likely to have a mean between 50.32 mm and 50.48 mm?

   Hint: Find the probability from the area under the normal curve between the two values given, and use this as a percentage of the total number (30×80) of shafts sampled.

6) Every day a company produces 600 packets of a chemical each having a mean mass of 15.900 kg with a standard deviation of 0.100 kg. Find the probability that the mass of a packet
   a) is greater than 16.100 kg   b) is less than 15.800 kg.

7) In the previous question if a random sample of 40 packets is taken from a daily output:
   a) estimate the limits of a sample mass for a confidence level of 97%,
   b) what confidence level would you expect for mass sample limits of 15.88 kg and 15.92 kg?

24 Regression and Correlation

Dependent and independent variables – the straight line – curve fitting – method of least squares – the least square line – regression – correlation – measures of correlation – the coefficient of correlation

INTRODUCTION

We have seen previously when plotting experimental data how the points may be 'scattered' but close enough to warrant drawing the 'best line' through them.

Generally we are concerned with the 'best *straight* line' as this is the important relationship if it can be established.

We have used our judgement in drawing this best fit line and for many purposes this is good enough. However there is a more scientific way of arriving at this line, and this is what the chapter is all about.

DEPENDENT AND INDEPENDENT VARIABLES

Consider the equation:

$$y = 3x + 2$$

We can give x any value we please and so calculate the corresponding value of y. Thus,

when
$$x = 0, \quad y = (3 \times 0) + 2 = 2$$
$$x = 1, \quad y = (3 \times 1) + 2 = 5$$
$$x = 2, \quad y = (3 \times 2) + 2 = 8 \quad \text{and so on}$$

The value of y therefore *depends* upon the value allocated to x. We therefore call y the *dependent variable*. Since we can give x any value we please x is called the *independent variable*. It is usual, when plotting a graph, to mark the values of the independent variable along the horizontal axis and this axis is frequently called the x-axis. Values of the dependent variable are then marked off along the vertical axis and this axis is often called the y-axis.

The Equation of a Straight Line

Remember the standard straight
line equation

$$y = mx + c$$

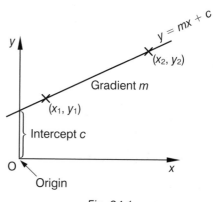

Fig. 24.1

The intercept on the y-axis is only
as shown in Fig. 24.1 if the origin O
is at the intersection of the axes.

Otherwise we use the two point
method shown below which works
whether the origin is present or not!

Two points (x_1, y_1) and (x_2, y_2), which
lie on the line, are chosen. They should be reasonably far apart, and
probably not given points which may lie off the line.

Since both these points lie on the line then their co-ordinates satisfy the

equation, giving the two equations $y_1 = mx_1 + c$
                        and $y_2 = mx_2 + c$

which may be solved simultaneously to find constants m and c.

Now just to try and confuse you, the equation of the straight line used in
this chapter has the gradient m replaced by b, and c replaced by a and
is written in different order as $y = a + bx$.

The reason for this is that formulae we shall use later are written based
on this arrangement. Everything we have learned up to now still works as
you will see from the next example.

EXAMPLE 24.1

The data for two variables x and y are given in this table:

x	2	5	7	10	14
y	8	17	23	32	44

Plot the graph, check if it is a straight line and using the form $y = a + bx$ find the values of a and b.

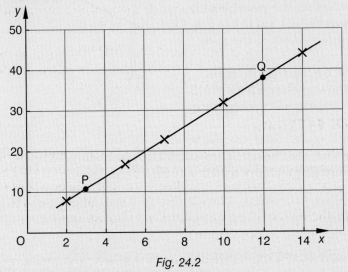

Fig. 24.2

The graph is shown plotted in Fig. 24.2 and it is seen to be a straight line. Hence the equation is of straight line form, say $y = a + bx$.

To find values of a and b we choose two points which lie on the line.

Now   P has co-ordinates $x = 3$ and $y = 11$
whilst   Q has co-ordinates $x = 12$ and $y = 38$

Now substituting these values in the straight line equation

Now for point P we have $11 = a + 3b$ [1]
 and for point Q we have $38 = a + 12b$ [2]

Now subtracting equation [1] from [2] then $27 = 9b$
∴ $b = 3$

Substituting $b = 3$ into equation [1] then $11 = a + 3 \times 3$
giving $a = 2$

Hence the straight line equation connecting x and y is $y = 2 + 3x$.

Interpolation

This means finding corresponding values of x and y *within the range of values* given in the original table, using the equation just found. For example when $x = 3.5$ then $y = 2 + 3 \times 3.5 = 10.5$

Extrapolation

As above but for values which lie outside the range of given values. Thus e.g. when $x = 15.5$ then $y = 2 + 3 \times 15.5 = 48.5$

Caution! It must be clearly understood that both the above techniques should only be used if the *graph is guaranteed* to be either a *straight line* or *a smooth curve* – especially true for extrapolation!!

CURVE FITTING

Readings which are obtained as a result of an experiment usually contain errors in measurement and observation. When the points are plotted on a graph it is usually possible to visualise a straight line or a curve which approximates to the data. Thus in Fig. 24.3 the data appears to be approximated by a straight line whilst in Fig. 24.4 the data is approximated by a smooth curve.

Figures 24.3 and 24.4 are called *scatter diagrams*. The problem is to find equations of curves or straight lines which approximately fit the plotted data. Finding equations for the approximating curves or straight lines is called *curve fitting*.

Individual judgement may be used to draw the approximating straight line or curve but this has the disadvantage that different individuals will obtain different straight lines or curves and hence different equations. To avoid this disadvantage the method of least squares is usually used to obtain the equation of the approximating curve or straight line.

In this chapter we shall confine out studies to straight lines.

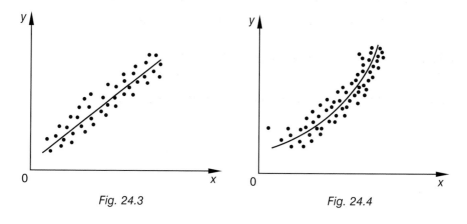

Fig. 24.3   Fig. 24.4

METHOD OF LEAST SQUARES

Consider Fig. 24.5 where an approximating straight line has been drawn to fit the given data. There is a deviation between the point (x_i, y_i) of the given data and the point (x_i, Y_i) which lies on the approximating straight line.

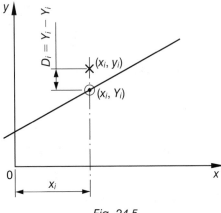

Fig. 24.5

This deviation is:

$$D_i = y_i - Y_i$$

The straight line having the property that:

$$\sum D_i^2 = D_1^2 + D_2^2 + D_3^2 + \ldots\ldots \text{ is a minimum}$$

is called the best fitting straight line.

THE LEAST SQUARE LINE

The best straight line approximating to a set of points (x_1, y_1), (x_2, y_2) ..., (x_n, y_n) has the equation

$$y = a + bx$$

It can be shown that

$$\sum y = aN + b \sum x \qquad [1]$$

$$\sum xy = a \sum x + b \sum x^2 \qquad [2]$$

These are a pair of simultaneous equations from which the constants a and b may be found, N being the number of points. The equation obtained is called the *least square line*. Only the least square line will be considered in this book.

EXAMPLE 24.2

Corresponding values obtained experimentally for two quantities x and y are:

x	2.25	12.0	22.0	31.5
y	4.5	6.0	7.5	9.0

x and y are connected by an equation of the type $y = a + bx$.
By finding the least square line determine suitable values for a and b.

x	y	xy	x^2
2.25	4.5	10.1	5.1
12.0	6.0	72.0	144.0
22.0	7.5	165.0	484.0
31.5	9.0	283.5	992.3
67.75	27.0	530.6	1625.4

$\sum y = 27.0$, $\sum x = 67.75$ and $N = 4$

$\therefore$ $\qquad\qquad\qquad 27.0 = 4a + 67.75b$ [1]

$\sum xy = 530.6$, $\sum x = 67.75$ and $\sum x^2 = 1625.4$

$\therefore$ $\qquad\qquad\qquad 530.6 = 67.75a + 1625.4b$ [2]

Solving equations [1] and [2] gives:

$$a = 4.15 \text{ and } b = 0.154$$

The least square line is:

$$y = 4.15 + 0.154x$$

EXAMPLE 24.3

The table below shows the
results of an experiment to
establish the relationship

t	25	50	75	100	125	150
R	20.7	21.6	22.2	23.0	23.9	24.6

between the resistance of a conductor (R ohm) and its temperature ($t\,°C$).

a) Find the equation connecting R and t if it is of the type $R = a + bt$.
b) Estimate the value of R when $t = 121.1\,°C$.

a) Here R is the dependent variable
(corresponding to y) and t is the
independent variable (corresponding
to x).

$$\sum R = 136, \ \sum t = 525 \text{ and } N = 6$$

$$136 = 6a + 525b$$

t	R	tR	t^2
25	20.7	517.5	625
50	21.6	1080.0	2500
75	22.2	1665.0	5625
100	23.0	2300.0	10 000
125	23.9	2987.5	15 625
150	24.6	3690.0	22 500
525	136.0	12 240.0	56 875

$$\sum tR = 12\,240, \ \sum t = 525 \text{ and } \sum t^2 = 56\,875$$

$$12\,240 = 525a + 56\,875b$$

Solving these two equations gives:

$$a = 19.95 \text{ and } b = 0.031$$

Hence $\qquad\qquad\qquad R = 19.95 + 0.031t$

b) When $\qquad t = 121.1\,°C,$

$$R = 19.95 + (0.031 \times 121.1) = 23.70\,\text{ohm}$$

REGRESSION

Suppose we are given several corresponding values of x and y which, when plotted, approximate to a straight line. To find a value for y corresponding to a stated value of x (which is not included in the given data) we first obtain the equation for the least square line which fits the given data. This line is called the *regression line of y on x* because y is estimated from x. (Note that the regression of y on x is determined by minimising sums of squares vertically.)

Sometimes we wish to estimate the value of x corresponding to a given value of y. In this case we use the regression line of x on y. This means interchanging the variables so that x is the dependent variable and y is the independent variable. (Note that the regression line of x on y is determined by minimising sums of squares horizontally.)

Generally the regression line of y on x is not the same as the regression line of x on y.

The regression line for y on x is:

$$y = a + bx$$

As shown previously the constants a and b may be found by solving the following pair of simultaneous equations:

$$\sum y = aN + b \sum x \qquad [1]$$

$$\sum xy = a \sum x + b \sum x^2 \qquad [2]$$

The regression line for x on y is:

$$x = a_1 + b_1 y$$

The constants a_1 and b_1 may be found by solving the following pair of simultaneous equations:

$$\sum x = a_1 N + b_1 \sum y \qquad [1]$$

$$\sum xy = a_1 \sum y + b_1 \sum y^2 \qquad [2]$$

EXAMPLE 24.4

The table below shows corresponding values of x and y obtained in an experiment:

x	20.7	21.0	21.3	21.7	22.0	22.3	22.7	23.0	23.3	23.7
y	22.0	22.1	21.7	22.7	21.7	22.7	23.0	22.7	22.8	23.7

a) Find the regression line for y on x.
b) Find the regression line for x on y.

x	y	xy	x^2	y^2
20.7	22.0	455.4	428.5	484.0
21.0	22.1	464.1	441.0	488.4
21.3	21.7	462.2	453.7	470.9
21.7	22.7	492.6	470.9	515.3
22.0	21.7	477.4	484.0	470.9
22.3	22.7	506.2	497.3	515.3
22.7	23.0	522.1	515.3	529.0
23.0	22.7	522.1	529.0	515.3
23.3	22.8	531.2	542.9	519.8
23.7	23.7	561.7	561.7	561.7
$\sum$ 221.7	225.5	4995.0	4924.3	5070.6

a)
$$225.1 = 10a + 221.7b \quad\quad [1]$$
$$4995.0 = 221.7a + 4924.3b \quad\quad [2]$$

Solving these equations gives $a = 11.60$ and $b = 0.492$.
Hence the regression line for y on x is:

$$y = 11.60 + 0.492x$$

b)
$$221.7 = 10a_1 + 225.1b_1 \quad\quad [1]$$
$$4995.0 = 225.1a_1 + 5070.6b_1 \quad\quad [2]$$

Solving these equations gives $a_1 = -6.51$ and $b_1 = 1.274$.
Hence the regression line for x on y is:

$$x = -6.51 + 1.274y$$

The two regression lines are drawn in Fig. 24.6 where it will be seen that there is an angle between them. Thus the regression line for y on x is not the same as the regression line for x on y.

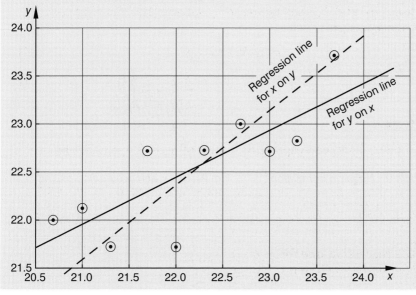

Fig. 24.6

CORRELATION

When corresponding values of two variables (x and y) obtained by experiment are plotted a scatter diagram like that shown in Fig. 24.3 is obtained. A rough relationship, or correlation, is seen to exist between x and y.

Correlation is closely associated with regression. We seek to determine how well a linear (or other) equation describes the relationship between the two variables. When the points on a scatter diagram are such that they approximate to a straight line, the correlation is said to be linear. Only linear correlation will be considered.

The correlation may be positive, precise or negative. For positive correlation (Fig. 24.7) large values of y accompany large values of x. For the correlation to be precise (Fig. 24.8) *all* the points on the scatter diagram must lie on a straight line. For negative correlation (Fig. 24.9), the values of y decrease as the values of x increase. If no relationship is indicated by the points on the scatter diagram we say that there is no correlation between x and y, i.e. x and y are uncorrelated (Fig. 24.10).

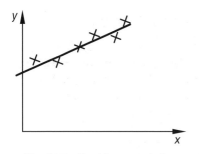

Fig. 24.7　Positive correlation

Fig. 24.8　Precise correlation

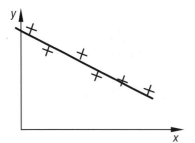

Fig. 24.9　Negative correlation

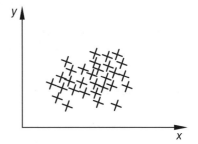

Fig. 24.10　x and y are uncorrelated

MEASURES OF CORRELATION

We have seen on page 318 that the least square regression line for y on x is given by the equation:

$$y = a + bx$$

The least square regression line for x on y is given by:

$$x = a_1 + b_1 y$$

These two equations are identical only if the correlation is precise, that is, only if all the points on the scatter diagram lie on a straight line.

We often can see, by direct observation of the scatter diagram, how well a straight line describes the relationship between two variables. In Fig. 24.7, for instance, we see that the straight line describes the relationship between x and y very well indeed. In Fig. 24.3, however, the relationship is not so well defined by the straight line.

Hence, in order to deal with the problem of scatter, we have a measure of correlation.

THE COEFFICIENT OF CORRELATION

For perfect correlation the regression lines of y on x and x on y coincide. When the lines do not coincide we need some quantity which will measure the degree of correlation. The quantity used is called the *coefficient of correlation* and it is represented by the symbol r. The value of r may be calculated from the formula

$$r = \frac{\Sigma xy - N\bar{x}\,\bar{y}}{N\,\sigma_x\,\sigma_y}$$

This is known as   *The product–moment formula for the linear correlation coefficient*

The example which follows shows how the formula is applied.

EXAMPLE 24.5

For the data of Example 24.4, calculate the value of r.

Since $\qquad \sum xy = 4995.0 \qquad \sum x = 221.7 \qquad \sum y = 225.1$

$\qquad\qquad \sum x^2 = 4924.3 \qquad \sum y^2 = 5070.6 \qquad$ and $\quad N = 10$

$$\bar{x} = \frac{\sum x}{N} = \frac{221.7}{10} = 22.17$$

$$\bar{y} = \frac{\sum y}{N} = \frac{225.1}{10} = 22.51$$

$$\sigma_x = \sqrt{\frac{\sum x^2}{N} - \bar{x}^2} = \sqrt{\frac{4924.3}{10} - 22.17^2} = 0.960$$

$$\sigma_y = \sqrt{\frac{\sum y^2}{N} - \bar{y}^2} = \sqrt{\frac{5070.6}{10} - 22.51^2} = 0.600$$

$$r = \frac{4995.0 - 10 \times 22.17 \times 22.51}{10 \times 0.960 \times 0.600} = 0.787$$

An Alternative Method to Find the Linear Correlation Coefficient

If the equations for the regression lines of y on x and x on y are known, then the following formula may be used:

$$\boxed{r = \sqrt{b\, b_1}}$$

EXAMPLE 24.6

As in the previous example using the data from Example 24.4 we have

the regression line of y on x is $y = 11.60 + 0.492x$

and regression line of x on y is $x = -6.51 + 1.274y$

So $\qquad\qquad r = \sqrt{0.492 \times 1.274} = 0.792$

Use of the r Value

The values of r obtained by the two methods are only approximate, but sufficiently accurate to make the following judgements:

for precise positive correlation, the value of r is $+1$
for precise negative correlation, the value of r is -1

If the value of r is near to zero it means that there is practically no correlation between the variables.

Exercise 24.1

1) The table opposite gives corresponding values of x and y.

x	1	2	3	4	5
y	8	11	14	17	20

Show that x and y are related by an equation of the type $y = a + bx$ and find values for a and b.

2) The following table gives values of x and y which are connected by an equation of the type $y = a + bx$. Find suitable values for a and b.

x	2	4	6	8	10	12
y	10	16	22	28	34	40

3) A test on a metal-filament lamp gave the following values of resistance (R ohm) at various voltages (V volt):

V	62	75	89	100	120
R	100	118	136	149	176

R and V are connected by an equation of the type $R = a + bV$. Determine suitable values for a and b by finding the least square line.

4) During a test with a thermocouple pyrometer the e.m.f. (E millivolt) was measured against the temperature of the hot junction ($t\,°C$) and the following results obtained:

t	200	300	400	500	600	700	800	900	1000
E	6	9.1	12.0	14.8	18.2	21.0	24.1	26.8	30.2

E and t are connected by an equation of the type $E = a + bt$. By finding the least square line find values for a and b. Hence find the value of E when $t = 840\,°C$.

5) Find:
a) the regression line for y on x,
b) the regression line for x on y for the values given in the table:

x	62	75	89	100
y	100	117	135	149

6) The regression line of y on x is $y = 24.0 + 1.2x$ and the regression line of x on y is $y = -12.0 + 0.7y$. Calculate the value of the linear correlation coefficient.

7) Corresponding values of x and y are given in the following table:

x	20	40	60	80
y	101	108.1	115.1	122

Calculate the product-moment coefficient of linear correlation.

8) The table opposite shows the distance travelled (D km) and the time taken (T days) by 8 lorries.

D	1320	344	1712	880	768	1472	2160	520
T	3.5	1.0	4.0	2.0	1.0	3.0	4.5	1.5

a) Find the regression line for D on T.
b) Find the regression line for T on D.
c) What is the value of the correlation coefficient?

Answers to Exercises

Exercise 1.1

1) $a = 70$, $b = 50$

2) $k_{\mathrm{m}} = 0.016 + \dfrac{0.023}{u}$

3) $pv = 0.000\ 667$

4) $k = 0.2$

5) $a = 0.761$, $b = 10.1$

6) $m = 0.040$, $c = 0.20$

7) $m = 0.1$, $c = 1.4$

Exercise 1.2

1) $a = 3$, $n = 0.5$

2) $n = 4.05$, for $I = 20$ read $I = 35$

3) $t = 0.3m^{1.5}$

4) $k = 100$, $n = -1.2$

5) $a = 245$, $b = 33$

6) $\mu = 0.5$, $k = 5$

7) $I = 0.02$, $T = 0.2$

8) $k = 23.3$, $c = 2.99$

9) $V = 100$, $t = 0.0025$

Exercise 2.1

1) 1, 2

2) 4, 5

3) 18, 28

4) 5400, 1700, £5400, £3400

5) 25 Ω, 0.005, 31.25 Ω

6) £250, £6500

7) £14 000, £16 000

8) $a = 0.4$, $b = 50$, 170 N

Exercise 2.2

1) $x = -1$, $y = 1$; $x = 4$, $y = 6$

2) $x = 0$, $y = 5$; $x = 3$, $y = 11$

3) $x = 1$, $y = 3$; $x = -0.2$, $y = -3$

4) $x = 2.39$, $y = 6.91$; $x = 0.26$, $y = 0.54$

5) 60 m × 80 m

Exercise 2.3

1) 3, 4

2) 4 repeated

3) +3, −3

4) 3, −5

5) 0.667, 7

6) −5, −1.5

7) 2.414, −0.414

8) 2.181, 0.153

9) +0.745, −0.745

Exercise 2.4

1) 1, −1, 4

2) 2, 0.33, −1.20

3) 1

4) −1, 1 repeated

5) 3, 1.18, −0.43

6) 2

7) 1, 0.57, −2.91

8) −3, 2 repeated

9) 2, 0.21, −1.35

10) −1

11) 7.9 m

12) 3.32 m

13) 5 m

14) 3

Exercise 3.1

1) a) 2×2 **b)** 2×1
 c) 3×3 **d)** 2×4

2) a) 9   **b)** 4   **c)** n^2

3) a) $\begin{pmatrix} 1 & 3 \\ 2 & 4 \end{pmatrix}$ **b)** $(5 \quad -6)$

 c) $\begin{pmatrix} a & 2 & x \\ b & 3 & -6 \\ 4 & 5 & 0 \end{pmatrix}$ **d)** $\begin{pmatrix} 1 & 6 \\ -2 & 2 \\ -3 & 0 \\ -4 & -1 \end{pmatrix}$

4) a) $\begin{pmatrix} 0 & 0 \\ 9 & 2 \end{pmatrix}$ **b)** $\begin{pmatrix} 4 & 2 \\ -3 & 2 \end{pmatrix}$

 c) $\begin{pmatrix} 5/6 & 1/2 \\ 5/6 & 1 \end{pmatrix}$

5) $a = -8$, $b = 5$, $c = 1$

6) $\begin{pmatrix} 1/3 & 1/20 \\ 1/30 & 1/18 \end{pmatrix}$

7) $\begin{pmatrix} 5 & 8 \\ 12 & -2 \end{pmatrix}$ **8)** $\begin{pmatrix} 0 \\ 5 \end{pmatrix}$

Exercise 3.2

1) a) $\begin{pmatrix} 6 & 0 \\ -4 & 2 \end{pmatrix}$ **b)** $\begin{pmatrix} -12 & 3 \\ 9 & -6 \end{pmatrix}$

 c) $\begin{pmatrix} -6 & 3 \\ 5 & -4 \end{pmatrix}$ **d)** $\begin{pmatrix} 18 & -3 \\ -13 & 8 \end{pmatrix}$

2) a) $\begin{pmatrix} 14 & 0 \\ 8 & -2 \end{pmatrix}$ **b)** $\begin{pmatrix} 2 & 1 \\ 3 & 1 \end{pmatrix}$

c) $\begin{pmatrix} 5 & 11 \\ 10 & 22 \end{pmatrix}$ **d)** $\begin{pmatrix} a & b \\ c & d \end{pmatrix}$

e) $\begin{pmatrix} ka & kb \\ kc & kd \end{pmatrix}$

3) a) $\begin{pmatrix} 7 & 10 \\ 15 & 22 \end{pmatrix}$ **b)** $\begin{pmatrix} 3 & -5 \\ 5 & 8 \end{pmatrix}$

c) $\begin{pmatrix} 8 & 10 \\ 20 & 18 \end{pmatrix}$ **d)** $\begin{pmatrix} 18 & 15 \\ 40 & 48 \end{pmatrix}$

e) $\begin{pmatrix} 13 & 10 \\ 40 & 53 \end{pmatrix}$

Exercise 3.3

1) a) 24   **b)** −6   **c)** 14
2) a) $x = 1$, $y = 2$
  b) $x = 4$, $y = 3$
  c) $x = 0.5$, $y = 0.75$

Exercise 3.4

1) $\dfrac{1}{3}\begin{pmatrix} 4 & -5 \\ -1 & 2 \end{pmatrix}$

2) $\begin{pmatrix} 3 & -5 \\ -1 & 2 \end{pmatrix}$

3) $\dfrac{1}{4}\begin{pmatrix} 2 & -2 \\ -1 & 3 \end{pmatrix}$

4) No inverse

5) $\dfrac{1}{320}\begin{pmatrix} 4 & -24 \\ -24 & 224 \end{pmatrix}$

6) No inverse

7) $\dfrac{1}{7}\begin{pmatrix} 2 & -3 \\ 1 & 2 \end{pmatrix}$

8) $\dfrac{1}{13}\begin{pmatrix} 5 & 3 \\ -1 & 2 \end{pmatrix}$

9) $\begin{pmatrix} 1 & -1 \\ 0 & 1 \end{pmatrix}$

10) a) $\dfrac{1}{2}\begin{pmatrix} 2 & 0 \\ -3 & 1 \end{pmatrix}$ **b)** $\begin{pmatrix} 2 & -5 \\ -1 & 3 \end{pmatrix}$

c) $\dfrac{1}{2}\begin{pmatrix} 19 & -5 \\ -11 & 3 \end{pmatrix}$ **d)** $\begin{pmatrix} 3 & 5 \\ 11 & 19 \end{pmatrix}$

e) $\dfrac{1}{2}\begin{pmatrix} 19 & -5 \\ -11 & 3 \end{pmatrix}$ **f)** equal

Exercise 3.5

1) $x = 6$, $y = -5$
2) $x = 1$, $y = 5$
3) $x = 3$, $y = -1$

4) $x = 2$, $y = -3$
5) $x = 16/11$, $y = 9/11$
6) $x = 2$, $y = -5$

Exercise 4.1

1) 33, 60
2) 23rd
3) a) 27   **b)** 203
4) 3.75, 4.20, 4.65, 5.10, 5.55, 6.00
5) −10, 2, 14 or 6, $-8\frac{2}{3}$, $-23\frac{1}{3}$
6) Nine
7) £274 750
8) 88.3 hours
9) a) 720 m
  b) 135 m
  c) 24.5 seconds
10) 40, 110, 179, 249, 318

Exercise 4.2

1) 1.949
2) −2187
3) 3/8, 1/8, 1/24
4) a) 118.1   **b)** 120
5) 3, 6, 12, 24, 48, 96
6) 6, 11, 22, 42, 81, 156 rev/min
7) 47, 78, 129, 213, 353, 584 rev/min
8) a) 255, 673, 1092, 1510, 1928, 2346, 2765, 3183 rev/min
  b) 255, 366, 525, 752, 1079, 1547, 2219, 3183 rev/min
9) a) £31 789   **b)** £120 000
10) a) 15.1   **b)** 333
11) a) 4877   **b)** 1706

Exercise 5.1

1) $1 + 5z + 10z^2 + 10z^3 + 5z^4 + z^5$
2) $p^6 + 6p^5q + 15p^4q^2 + 20p^3q^3$
  $+ 15p^2q^4 + 6pq^5 + q^6$
3) $x^4 - 12x^3y + 54x^2y^2 - 108xy^3 + 81y^4$
4) $32p^5 - 80p^4q + 80p^3q^2 - 40p^2q^3$
  $+ 10pq^4 - q^5$
5) $128x^7 + 448x^6y + 672x^5y^2$
  $+ 560x^4y^3 + 280x^3y^4$
  $+ 84x^2y^5 + 14xy^6 + y^7$
6) $x^3 + 3x + 3/x + 1/x^3$
7) $1 + 12x + 66x^2 + 220x^3 + ...$
8) $1 - 28x + 364x^2 - 2912x^3 + ...$
9) $p^{16} + 16p^{15}q + 120p^{14}q^2 + 560p^{13}q^3$
  $+ ...$
10) $1 + 30y + 405y^2 + 3240y^3 + ...$

11) $x^{18} - 27x^{16}y + 324x^{14}y^2 - 2268x^{12}y^3$
$+ \dots$

12) $x^{22} + 11x^{18} + 55x^{14} + 165x^{10} + \dots$

13) $35a^4b^3$

14) $210a^4b^6$

15) $43\,758a^{10}b^8$

Exercise 6.1

1) **a)** $18 + j10$ **b)** $-j$
   c) $8 - j10$

2) **a)** $-1 + j3$ **b)** $-4 - j3$
   c) $10 - j3$

3) **a)** $-9 + j21$
   b) $-36 - j32$
   c) $-9 + j40$
   d) 34
   e) $-21 - j20$
   f) $18 - j30$
   g) $0.069 - j0.172$
   h) $-0.724 + j0.690$
   i) $-0.138 - j0.655$
   j) $0.644 + j0.616$
   k) $3.5 - j0.5$
   l) $0.2 + j0.6$

4) **a)** $1, j0.5$ **b)** $0, j5$ **c)** $-3, j3$

5) **a)** $-1 \pm j$ **b)** $\pm j3$

6) $0.634 - j0.293$

7) **a)** $3 + j4$ **b)** $0.4 + j0.533$
   c) $0.692 + j2.538$

Exercise 6.2

2) Mod 5, Arg 53.13°;
Mod 5, Arg $-36.87°$

3) 3.61, 146.32°

4) 4.47, $-153.43°$

5) **a)** $5\underline{/36.87°}$
   b) $5\underline{/-53.13°}$
   c) $4.24\underline{/135°}$
   d) $2.24\underline{/-153.43°}$
   e) $4\underline{/90°}$
   f) $3.5\underline{/-90°}$

6) **a)** $2.12 + j2.12$ **b)** $-4.49 + j2.19$
   c) $4.32 - j1.57$ **d)** $-1.60 - j2.77$

7) **a)** 14.62°   **b)** 345 watts

Exercise 7.1

1) 0

2) 0

3) 0

4) 1

5) 1

6) 1

7) 1

8) 1

9) 1

10 1

11) 0

12) 1

13)   (i)

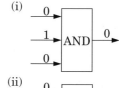

    (ii)

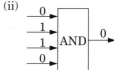

    (iii)

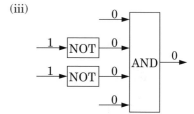

    (iv)

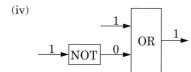

    (v)

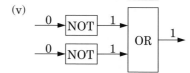

    (vi)

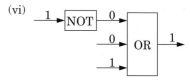

    (vii)

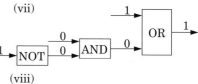

    (viii)

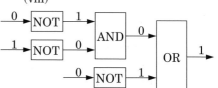

Exercise 7.2

No answers!!!

Exercise 7.3

1) a)

B\A	0	1
0		1
1		1

b)

B\A	0	1
0		1
1	1	1

c)

Y\X	0	1
0		1
1	1	

2 a)

BC\A	0	1
00	1	1
01	1	1
11	1	1
10		1

b)

BC\A	0	1
00	1	1
01		1
11		
10	1	1

c)

BC\A	0	1
00	1	
01		
11		
10	1	1

3) a)

YZ\WX	00	01	11	10
00				
01			1	1
11			1	1
10				

b)

CD\AB	00	01	11	10
00	1	1	1	
01	1	1	1	
11	1	1		1
10	1	1		

c)

YZ\WX	00	01	11	10
00	1			
01	1	1	1	1
11	1	1	1	1
10				

4 a)

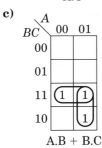

B\A	00	01
00		1
01	1	1

A + B

b)

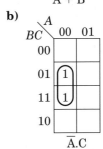

BC\A	00	01
00		
01	1	
11	1	
10		

$\overline{A}.C$

c)

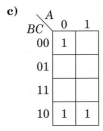

BC\A	00	01
00		
01		
11	1	1
10		1

A.B + B.C

d)

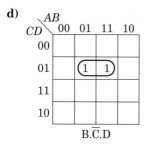

B.C̄.D

Exercise 7.4

1) a) 51   **b)** 143   **c)** 307

2) a) 35_8 **b)** 363_8 **c)** 3404_8

3) a) 715   **b)** 5614

5) a) 10111   **b)** 1111101
   c) 1100001

6) a) 22   **b)** 57   **c)** 90

7) a) 0.8125 **b)** 0.4375 **c)** 0.1875

8) a) 0.011   **b)** 0.0101 **c)** 0.111

9) a) 0.001 010 1
   b) 10010.01110 11
   c) 1 101 100.101 101 0

10) a) 1110
    b) 100 110
    c) 111 000

11 a) 100 011
    b) 1 101 110
    c) 101 011 111

Exercise 8.1

1) a) (8.60, 54.5°)
   b) (3.61, 123.7°)
   c) (3.61, 303.7°)
   d) (5, 233.1°)

2) a) (1.64, 1.15)
   b) (−1.81, 2.39)
   c) (−0.75, −1.30)
   d) (0.401, −0.446)
   e) (2.15, −0.824)
   f) (3.21, −3.83)

3)

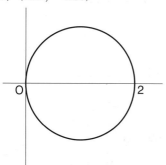

4)

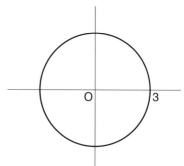

5)

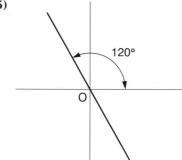

6)

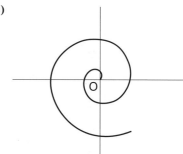

7)

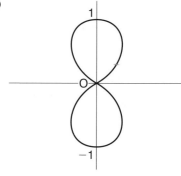

8)

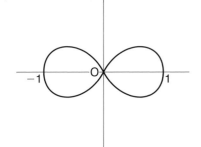

9)

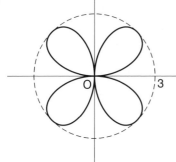

10)

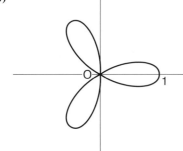

11)

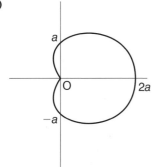

Exercise 9.1

1) 45.79 mm
2) 20.90 mm
3) 1.64 mm

4) 1.52°, 13.04 mm, 9.59 mm
5) 65.77°, 29.71 mm
6) 53.01 mm
7) 31.99 mm
8) 4.4°, 25.51 mm
9) 104.98 mm
10) 5.18 mm
11) 2.887 mm, 53.44 mm
12) 30.53 mm

Exercise 9.2

1) 2408 mm
2) 5369 mm
3) 12.63 m
4) 1287 mm
5) 2971 mm
6) 5740 mm
7) 16.01 m
8) 2215 mm
9) 36.1 m
10) BD = DF = AC = CE = 2.66 m
    GB = FH = 4.41 m
    BC = CF = 3.59 m
    GA = EH = 3.70 m

Exercise 10.1

1) 9.17 m, 24.20°
2) 257.3 mm, 97.03°
3) 111.5 mm, 77.10°
4) 26.17 mm
5) 144.5 mm
6) 4.78 m, 4.09 m
7) 13.2 m^2
8) 9330 mm^2
9) 4105 mm^2
10) 576 mm^2

Exercise 11.1

1)

1.50 N
2.60 N

2)

2.57 m/s
3.06 m/s

3)

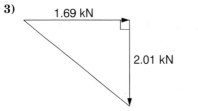

4)

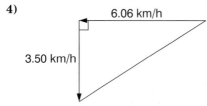

5)

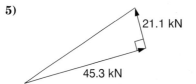

6) 5.00 kN down the slope,
8.66 kN at right angles to slope
7) 81.6 N horizontal,
40.1 N vertical
8) 1.93 m/s
9) 58.3 N, 31°
10) 703.5 km/h, 5.71° east of north
11) 0.568 m/s, 38.4° west of north
12) 353 N, anticlockwise 11.0°
13) 347 N along the track
67.4 N between wheel flanges and
rail
14) 3.61 V, anticlockwise 33.7°
15) 4.78 A, clockwise 13.9°
16) 10.1 V, clockwise 40.8°
17) 6.51 A, clockwise 33.0°
18) F_{AB} 76 kN tensile
F_{CB} 143 kN compressive

Exercise 12.1

No answers

Exercise 12.2

4) a) $\sin\left(\omega t + \dfrac{\pi}{2}\right)$ **b)** $\sin(\omega t - \pi)$

c) $\sin \omega t$ **d)** $\sin\left(\omega t - \dfrac{\pi}{6}\right)$

6) 3.61, 33.7°
7) $v_R = 7.7\sin(\theta - 19°)$
8) $i_R = 6.6\sin(\theta + 12°)$
9) $v_R = 51\sin(\theta - 11°)$

Exercise 13.1

1) $15x^2 + 14x - 1$
2) $3.5t^{-0.5} - 1.8t^{-0.7}$
3) 1.25
4) 1
5) 5, −3

Exercise 13.2

1) $6(3x + 1)$
2) $-15(2 - 5x)^2$
3) $-2(1 - 4x)^{-1/2}$
4) $-7.5(2 - 5x)^{1/2}$
5) $-4(4x^2 + 3)^{-2}$
6) $3\cos(3x + 4)$
7) $5\sin(2 - 5x)$
8) $8\sin 4x \cos 4x$
9) $21(\sin 7x)/\cos^4 7x$

10) $2\cos\left(2x + \dfrac{\pi}{2}\right)$

11) $-3(\sin x)\cos^2 x$

12) $-\dfrac{\cos x}{\sin^2 x}$

13) $\dfrac{1}{x}$

14) $-\dfrac{9}{x}$

15) $\dfrac{1}{2(2x - 7)}$

16) $-\dfrac{1}{e^x}$

17) $6e^{(3x + 4)}$
18) $8e^{(8x - 2)}$
19) $\frac{2}{3}(1 - 2t)^{-4/3}$
20) $\frac{3}{4}\cos(\frac{3}{4}\theta - \pi)$
21) $\{-\sin(\pi - \phi)\}/\cos^2(\pi - \phi)$

22) $-\dfrac{1}{2x}$

23) $Bke^{(kt - b)}$

24) $-\dfrac{1}{3}e^{(1 - x)/3}$

Exercise 13.3

1) a) $\sin x + x\cos x$
b) $e^x(\tan x + \sec^2 x)$
c) $1 + \log_e x$
2) $\cos^2 t - \sin^2 t$
3) $2(\tan\theta)\cos 2\theta + (\sec^2\theta)\sin 2\theta$
4) $e^{4m}(4\cos 3m - 3\sin 3m)$
5) $3x(1 + 2\log_e x)$
6) $6e^{3t}(3t^2 + 2t - 3)$
7) $1 - 3z + (1 - 6z)\log_e z$

8) a) $\dfrac{1}{(1-x)^2}$

   b) $\dfrac{1 - 2\log_e x}{x^3}$

   c) $\dfrac{e^x(\sin 2x - 2\cos 2x)}{\sin^2 2x}$

9) $\dfrac{11}{(3 - 4z)^2}$

10) $-\dfrac{2(\sin 2t + \cos\ 2t)}{e^{2t}}$

11) $-\text{cosec}^2\,\theta$

Exercise 14.1

1) 42 m/s

2) 6 m/s^2

3) a) 6 m/s

   b) 2.41 or -0.41 s

   c) 6 m/s^2

   d) 1 s

4) -0.074 m/s, 0.074 m/s^2

5) 10 m/s, 30 m/s

6) 3.46 m/s

7) a) 4 rad/s

   b) 36 rad/s^2

   c) 0 s or 1 s

8) a) -2.97 rad/s   **b)** 0.280 s

   c) -8.98 rad/s^2 **d)** 1.57 s

9) 62.5 kJ

Exercise 14.2

1) a) $T = 631$ **b)** $s = 1.12$

2) 4.34 mA per second

3) a) 0.18 seconds

   b) 1200 V per second

4) 10 years

Exercise 14.3

1) a) 11(max), -16(min)

   b) 4(max), 0(min)

   c) 0(min), 32(max)

2) a) 54   **b)** 2.5   **c)** $x = -2$

3) $(3, -15)$, $(-1, 17)$

4) a) -2 **b)** 1   **c)** 9

5) a) 12   **b)** 12.48

6) 10 V(max), 4 V(min)

7) 8

8) 12.5 m/s, 1.23 kW

9) 15 mm

10) 10 m

11) 4

12) 108 000 mm^3

13) 86 mm diam., 86 mm long

14) 405 mm

15) 28.9 mm diam., 14.4 mm high

Exercise 15.1

1) $\frac{5}{3}x^3 + x^2 + \dfrac{4}{x} + c$

2) $\frac{2}{3}x^{3/2} + 2x^{1/2} + c$

3) $-3\left(\cos\dfrac{x}{3}\right) + c$

4) $\frac{5}{3}(\sin 3\theta) + c$

5) $\phi - \frac{3}{2}(\cos\frac{2}{3}\phi) + c$

6) $2\left(\sin\dfrac{\theta}{2}\right) + \frac{2}{3}\left(\cos\dfrac{3\theta}{2}\right) + c$

7) $t^2 - \frac{1}{2}(\cos 2t) + c$

8) $\frac{1}{3}e^{3x} + c$

9) $-2e^{-0.5u} + c$

10) $\frac{3}{2}e^{2t} - 2e^t + c$

11) $-2x^{-x/2} + \frac{2}{3}e^{3x/2} + c$

12) $(\tan x) + (\log_e x) + c$

13) 7.75

14) 4.67

15) 0.561

16) 0

17) 0.521

18) 0

19) 0.586

20) 1.12

21) 1.33

22) 1.72

23) 0.0585

24) 0.253

25) 4.15

26) 51.1

27) 0.811

Exercise 15.2

1) $\frac{1}{8}(2x + 1)^4 + c$

2) $\frac{2}{3}(x + 3)^{3/2} + c$

3) $-\frac{1}{2}\log_e(1 - 2x) + c$

4) $-\frac{1}{8}(3 - x^2)^4 + c$

5) $-\frac{1}{2}\cos(2\theta - 1) + c$

6) $-\frac{1}{2}e^{(1 - 2x)} + c$

7) 0.693

8) 27.7

9) 0.0855

10) $\frac{1}{3}$

11) 4024

12) 0.3574

13) 0.25
14) 0.333
15) 2.054
16) $\frac{2}{3}$
17) 0.3133
18) -0.2152
19) 1.228
20) $\frac{4}{3}$
21) 0.3466

Exercise 15.3

1) $xe^x - e^x + c$
2) $\frac{3}{4}x^2(2\log_e 5x - 1) + c$
3) $\frac{2}{3}xe^{3x} - \frac{2}{9}e^{3x} + c$
4) 0.571
5) 2.142
6) 0.386
7) 0.265
8) 2
9) 0.494

Exercise 16.1

1) 136
2) 87
3) 5.167
4) 0.667
5) 3.75
6) 0
7) 4
8) 2
9) 6.39
10) 5.21

Exercise 16.2

1) a) 0.637 V   **b)** 0
2) a) 3.82   **b)** 0
3) 1 volt
4) 0 volt
5) 5 volt
6) 1.67 volt
7) 2.25 volt
8) a) 0.707 V   **b)** 0.707 V
9) a) 4.24   **b)** 4.24
10) 1.42, 2, 5.77, 2.36, 2.45 volts

Exercise 17.1

1) $\bar{x} = 38.7$ mm, $\bar{y} = 41.8$ mm
2) $\bar{x} = 68.3$ mm, $\bar{y} = 73.8$ mm
3) $\bar{x} = 19.1$ mm, $\bar{y} = 28.3$ mm
4) $\bar{x} = 50.0$ mm, $\bar{y} = 44.6$ mm
5) $\bar{y} = 29.7$ mm

Exercise 17.2

1) 57.4
2) 402
3) 0.0761
4) 171
5) 16.8
6) 0.105
7) a) 262 000 mm^3 **b)** 58 700 mm^3
8) 14.7 litres
9) 23 100 mm^3

Exercise 17.3

1) $\frac{1}{2}L$
2) $\frac{3}{4}H$ from apex
3) 0.911 m from small flat surface
4) 1.33 m from large flat surface
5) 0.688 from flat surface

Exercise 18.1

1) $I_{XX} = 1406$ cm^4, $I_{YY} = 354$ cm^4
2) $I_{XX} = 856$ cm^4, $I_{YY} = 1184$ cm^4
3) $I_{XX} = 603$ cm^4, $I_{YY} = 363$ cm^4
4) $I_{XX} = 459$ cm^4, $I_{YY} = 419$ cm^4

Exercise 18.2

1) $I_{XX} = 493$ cm^4, $I_{YY} = 173$ cm^4
2) $I_{XX} = 523$ cm^4, $I_{YY} = 1100$ cm^4
3) $I_{XX} = 136$ cm^4, $I_{YY} = 60$ cm^4
4) $I_{XX} = 375$ cm^4, $I_{YY} = 266$ cm^4
5) $I_{XX} = 133$ cm^4, $I_{YY} = 227$ cm^4
6) $I_{XX} = 274$ cm^4, $I_{YY} = 219$ cm^4

Exercise 18.3

1) a) 258 cm^4
   b) 517 cm^4
   c) 2520 cm^4
2) a) $I_{XX} = I_{YY} = 168$ cm, $J = 336$ cm^4
   b) $k_{XX} = 2.34$ cm
3) a) 1600 cm^4
   b) 1820 cm^4
   c) 909 cm^4

Exercise 19.1

1) 0.0133 kg m^2
2) 104 mm
3) 21.4 kg m^2
4) a) 8390 kg m^2 **b)** 32 300 kg m^2
5) a) 288 kg m^2 **b)** 1050 kg m^2
6) 17 700 kg m^2
7) 7.28 kg m^2

Exercise 20.1

1) **a)** 6.89 **b)** 7.12 **c)** 7.00
2) **a)** 0.851 **b)** 0.846 **c)** 0.848
3) **a)** 0.386 **b)** 0.387 **c)** 0.386
4) **a)** 0.368 **b)** 0.365 **c)** 0.366
5) **a)** 0.507 **b)** 0.495 **c)** 0.499
6) **a)** 0.0176 **b)** 0.0169 **c)** 0.0171
7) **a)** 0.647 **b)** 0.647 **c)** 0.647
8) Both 10 and 12 intervals give 2.42

Exercise 21.1

1) $\frac{1}{13}$
2) $\frac{1}{9}$
3) $\frac{5}{13}$
4) **a)** $\frac{2}{5}$ **b)** $\frac{1}{4}$ **c)** $\frac{3}{5}$
   d) $\frac{17}{20}$ **e)** $\frac{1}{5}$
5) $\frac{1}{2}$
6) **a)** 2% **b)** 0.02
7) 0.9

Exercise 21.2

1) $\frac{3}{13}$
2) 0.45
3) 0.5
4) $\frac{1}{3}$
5) $\frac{1}{36}$
6) **a)** (i) $\frac{1}{8}$ (ii) $\frac{3}{8}$
   b) (i) $\frac{1}{24}$ (ii) $\frac{1}{12}$
7) **a)** $\frac{1}{4}$ **b)** $\frac{1}{8}$ **c)** $\frac{7}{8}$ **d)** $\frac{1}{2}$
8) **a)** $\frac{8}{27}$ **b)** $\frac{1}{27}$ **c)** $\frac{2}{9}$ **d)** $\frac{4}{9}$
9) **a)** $\frac{4}{27}$ **b)** $\frac{5}{27}$
10) **a)** $\frac{1}{3}$ **b)** $\frac{5}{12}$ **c)** $\frac{1}{24}$ **d)** $\frac{1}{12}$
11) **a)** 0.32 **b)** 0.03
    c) (i) 0.0196 (ii) 0.04
    (iii) 0.1652 (iv) 0.1472
12) **a)** $\frac{15}{92}$ **b)** $\frac{91}{276}$ **c)** $\frac{35}{138}$ **d)** $\frac{35}{138}$

Exercise 21.3

1) 0.6561, 0.2916, 0.0486, 0.0036, 0.0001
2) **a)** 0.3585
   b) 0.2641
3) 0.1916
4) **a)** 0.3632
   b) 0.3725
   c) 0.2643
5) 0.1108
6) 0.3232

Exercise 21.4

1) 0.6703, 0.2681, 0.0536, 0.0072, 0.0007
2) 0.2240
3) **a)** 0.2707
   b) 0.5940
4) 0.4422
5) 56.65%
6) **a)** 0.0498
   b) 0.1992
   c) 0.5768
7) 0.7769
8) 0.4060

Exercise 22.1

1) 10.81%, 1.2782%
2) 11.4925 N/mm^2
   0.0145 2 N/mm^2
3) 99.93 W, 0.17 W
4) 12.67, 2.98
5) 196, 22.8 litres
6) **a)** 18.54 mm and 18.66 mm
   b) (i) 16% (ii) 2%
7) 1920

Exercise 22.2

1) **a)** -1.6 **b)** -0.4
   c) 1.0 **d)** 2.2
2) **a)** 0.3849 **b)** 0.2734
   c) 0.4865 **d)** 0.0487
   e) 0.2730 **f)** 0.2148
   g) 0.0174 **h)** 0.8051
   i) 0.8686 **j)** 0.9936
3) **a)** 18.60 ± 0.06 mm
   b) (i) 2.3% (ii) 0.6% (iii) 53.3%
4) **a)** 28.6% **b)** 66.8% **c)** 88.8%
   d) 2.3% **e)** 6.1%
5) 1810
6) $\bar{x} = 10.000$, $\sigma = 0.0128$
   a) 2.5% **b)** 0.3% **c)** 75.8%
7) $\bar{x} = 1.6847$ m, $\sigma = 0.0956$ m, 31.8%
8) $\bar{x} = 170.0$, $\sigma = 0.763$ kg, 89%
9) $\bar{x} = 12.50$, $\sigma = 0.0112$, 598
10) 115, 673
11) 95
12) $\bar{x} = 20.00\ \mu$F, $\sigma = 0.0129\ \mu$F
13) $\bar{x} = 170.025$ kg, $\sigma = 0.765$ kg
14) 67.4%
15) 109
16) 6.5%
17) 48.0%

Exercise 23.1

1) a) 8
   b) 4.472
   c) 8
   d) 3.162
2) a) 635 g and 0.233 g
   b) 635 g and 0.232 g
3) 1.727 m and 0.0513 m
4) a) 0.709
   b) 0.0119
   c) 0.0033
5) a) 50.42 mm and 0.225 mm
   b) 50.42 mm and 0.223 mm
   c) 22
6) a) 0.023   b) 0.159

7) a) 15.867 kg and 15.933 kg
   b) 80%

Exercise 24.1

1) $y = 5 + 3x$
2) $y = 4 + 3x$
3) $R = 20.0 + 1.30V$
4) $E = -0.0078 + 0.030t$, 25.2 mV
5) a) $y = 20.2 + 1.29x$
   b) $x = -15.6 + 0.776y$
6) 0.92
7) Virtually unity
8) a) $D = 33.9 + 434T$
   b) $T = 0.15 + 0.0021D$
   c) 0.95

Index